LASERS
Harnessing the Atom's Light

James P. Harbison

Robert E. Nahory

**SCIENTIFIC
AMERICAN
LIBRARY**

A division of HPHLP
New York

Cover and text designer: Diana Blume

Harbison, James P.
     Lasers: harnessing the atom's light / James P. Harbison, Robert
E. Nahory.
          p.   cm.
     Includes bibliographical references and index.
     ISBN 0-7167-5081-3
     1. Lasers.   I. Nahory, Robert E.   II. Title.
QC688.H37   1997
621.36'6 — dc21
                                                                    97-40517
                                                                        CIP

ISSN 1040-3213

© 1998 by Scientific American Library

Printed in the United States of America

Scientific American Library
A division of HPHLP
New York

Distributed by W. H. Freeman and Company,
41 Madison Avenue, New York, NY 10010
Houndmills, Basingstoke RG21 6XS, England

First printing, 1997

This book is number 66 of a series.

Contents

To Our Families

Preface

"The conditions of modern life are changing at an astonishing rate." These words, taken from the 1922 edition of *Practical Physics* by Nobel laureate Robert A. Millikan, Henry G. Gale, and Willard R. Pyle, were true in the early part of this century, when the automobile and airplane were new, and remain true today. Scientists and engineers are now creating new technologies whose effects may be as far reaching in the century to come as those of Millikan's planes and cars. An example is the technology of global communications. Optical fiber systems are under construction that will support the exchange of information worldwide at unprecedented rates.

The engine that drives these optical communications systems is the laser, a relatively simple device that harnesses the atom's light to form a directed, highly controlled beam. A laser beam has a number of specialized properties that make it an ideal light source for a host of sophisticated applications. The invention and development of the laser, along with the acquisition of the requisite basic understanding, required a large fraction of the twentieth century. Indeed, research to develop ever more sophisticated lasers continues today.

The authors of this book were privileged to play a part in the development of these devices. Although not involved in the original laser work of the late 1950s and 1960s, we were in the right place at the right time to help build upon it. One of us (James P. Harbison) was captivated by this dazzling new technology while still in high school in the late 1960s. Inspired by one of C. L. Stong's "Amateur Scientist" columns in *Scientific American,* he and a fellow classmate attempted to build their own helium-neon laser. Although they could never make their device lase, the quest led Jim to a lifelong fascination with physics and, eventually, to a place on the team that created, some twenty years later, some of the world's smallest lasers at the cutting edge of research in the field.

The environments where we both chose to practice our physics, the Bell Labs of the 1970s and 1980s and Bellcore in the late 1980s, were exciting and fertile grounds for laser research. In both labs, teams of researchers worked on the myriad technologies required for the development of a fully functioning optical fiber communications system. While the work was difficult and demanded a heavy commitment of time, we were inspired by our awareness of its place in history as, step by step, we saw one barrier after another overcome.

From time to time we would step back briefly from our immersion in this fast-paced quest and observe the process itself. As we watched optical fiber communications coming closer to reality we would be filled with a sense of wonder. In writing this book, we have again had the opportunity to step back, for a more extended length of time, to reflect on the recent development of lasers of all kinds. We have thereby achieved a much broader perspective than those close to the research fray are accustomed to having. Our observations from this new perspective have once again filled us with wonder at the achievements that workers in this field have made in a short period of time. In the process we have developed a renewed appreciation for the basic quantum physics that lies at the heart of the operation of lasers, and we have been gratified to see the thread of the key developments of twentieth-century physics directly connecting to the creation of this thoroughly novel and yet extremely practical device. We hope that our book provides the reader a similar opportunity to view some of the many facets of lasers: their many forms, their operating mechanisms, and their various applications. We will be satisfied if the reader, too, catches a glimpse of the wonder that the field inspires in us.

We are grateful to the many who made possible the writing of this book, including a panoply of people, too numerous to mention by name, back through our entire lives. We particularly remember inspirational leaders such as the late Dr. Rudolph Kompfner of Bell Labs, who foresaw the possibility of using lasers for communications before any of these devices even existed. Kompfner had the charming habit of showing up at unexpected times in one's lab or office, always a welcome visitor. Discussions with him never failed to leave us eager to go forward. The stimulation he provided, along with our interactions with an outstanding set of peer researchers, created a working life of enviable richness.

Also important to us were our collaborators in research. Among them we wish to acknowledge Martin A. Pollack, J. Christian DeWinter, Edgar D. Beebe, and Raymond J. Martin, long-term research partners of Robert E. Nahory, who all kept "progress progressing" and made the labor meaningful. James P. Harbison wishes to thank that high school classmate, Rich Balderston, and his supportive high school physics teachers Sam Tatnall and Wilbert Braxton, who helped "launch" him on his "laser career," and his parents Betty and Bob Harbison, who encouraged him to pursue that career despite the many doubters asking "What are you going to do with a physics degree?" In particular, he acknowledges the more recent professional support he received from his colleague Leigh Florez during their work in crystal growth for both the microlasers discussed in Chapter 7 and a host of other challenging projects along the way. Jim would also like to mention the critical role played by his research directors at Bell Labs and Bellcore, in particular by Venky Narayanamurti, who first convinced Jim to join Bell Labs and take up the fascinating art of molecular beam epitaxial crystal growth, and by Vassilis Keramidas who, as both colleague and friend, spurred him to push its limits into the unknown.

Robert E. Nahory thanks his parents, Elfrieda and Joseph Nahory, who knew the value of education. He gives special thanks to Miss Rankin, the third-grade teacher who introduced him to the evaporation of water. In so doing, she stimulated his lifelong studies on the physics of this phenomenon, which led him ultimately to the application of related principles in the growth of semiconductor crystals. And along with that, it led him and colleagues to the pioneering of new alloy materials that provided the first long-wavelength semiconductor laser custom-designed for fiber optic communications.

We are indebted to several colleagues for comments and advice on the manuscript and its illustrations. In particular, we thank Chris Palmstrøm for his uncommon effort to create the atomic crystal graphic at the opening of Chapter 6, Catherine Caneau for insights into specific semiconductor laser structures, Frans Spaepen for his help in understanding the various models of amorphous solids, and John Wullert for advice on several points. We owe thanks to our many colleagues who helped us to understand specialized aspects of lasers covered in the book. In particular, we thank Jim Allen for his help on free electron lasers and Andy Weiner for his elucidation of some of the fascinating aspects of ultrashort laser pulses. We also gratefully acknowledge the assistance of the Bellcore library staff, and in particular Cecelia Fiscus, Bob Arzberger, and Martha Broad, in locating many of the articles, books, and other background materials necessary to cover so wide a field.

Finally, we owe a great debt of gratitude to our friends at W. H. Freeman and Company, including Gary Carlson, who got us started on the book, and especially Jonathan Cobb, whose vision and constant encouragement led us to the creation of this book in its present form. Most of all we are indebted to Susan Moran, the outstanding editor who guided us through the process of creating a finite volume on a subject that we have lived with so closely for so many years. We are extremely grateful to the photo editor Larry Marcus for his expertise and diligence in finding the photographs that appear in these pages. To Tina Hastings we owe a large debt of gratitude for her expertise, and professional eye, in the assembly of the text and figures into final form for publication.

And to Diana Blume we express our appreciation for the lovely design seen in the interior pages. Everyone at Freeman with whom we have come in contact has been a pleasure to work with.

Above all, we are grateful to our wives and families, whose support and encouragement enabled us to put forth the effort required to prepare this book. Robert E. Nahory thanks his wife, Dawn, along with his children Jill, Doug, and Bob; James P. Harbison thanks his wife, Susan, and his three children Tom, Dan, and Betsy. The countless hours they've allowed us to spend on this labor are a true gift of the heart, and we thank them deeply for their patience, their love, and their support.

September 1997

LASERS

Laser beams directed by powerful telescopes have been used
for such diverse applications as measuring the atmospheric
disturbance created by air pollution and determining the
distance to the moon.

1

From Pure Physics to Pure Light

In the dark of a midsummer California night in 1969, in a room deep beneath the 120-inch Lick Telescope atop Mount Hamilton, a synthetic ruby crystal emits a beam of light powerful enough to reach the moon. A few days before, the first manned mission to the moon has left on its surface a mirror facing toward Earth. Reflected in this mirror, the beam of light from the ruby makes its half-million-mile round-trip journey at the speed of light to provide the means for measuring the distance to the moon to an accuracy far beyond what had ever been possible. The experiment provides a dramatic demonstration of the power of the laser, a device that emits light of exceptional intensity and uniformity.

In the observatory on that night, encircling the ruby in a twisted spiral, a powerful lamp emits a rapid series of intense flashes of light that penetrate the crystal. Atoms of chromium within the ruby crystal, evenly spread throughout the material, absorb the lamp's bright light. These chromium atoms, now poised in a high-energy excited state, one by one re-emit the absorbed energy, again in the form of light. Released in arbitrary directions, most of the light escapes out the side of the crystal, but a tiny portion remains trapped within by reflections. This tiny portion is the light emitted in a trajectory that follows precisely the central axis of the cylindrically shaped ruby crystal.

Now the avalanche of light we refer to as lasing begins. The trapped light is reflected back and forth down the ruby's cylindrical axis, returned upon itself over and over by the crystal's highly polished reflecting ends; at each pass the reflected light stimulates the emission of further light from the excited atoms within the crystal. The internal beam within the ruby builds in intensity until the crystal emits a brief but intense light beam out the end. The population of excited chromium atoms gives up its absorbed energy in the short time interval of only ten billionths of a second, but before then the power of the external beam has grown to a billion watts, equivalent to the output of tens of millions of conventional light bulbs! All the light waves in the pulse are launched in almost exactly the same direction, in what is referred to as a tightly collimated beam. The time the beam takes to complete its journey to the moon and back tells the waiting scientists the distance to the moon to within 15 meters.

The beam requires all the intensity the laser can muster to make it to the moon and back without becoming so dim as to be undetectable. The moon-ranging experiment depends on the laser's ability to store up the light energy generated over a longer time frame by the flash lamp and release it all at once in a single pulse. The result is that the intensity of this pulse is much greater than the intensity of the light from the lamp, since the same amount of light energy is concentrated much more tightly in time. In a sense, the lasing process is a way of transforming the light of the flash lamp, which could not meet the demanding requirements of the moon-ranging experiment, into a form of light that does.

The light produced by a laser has a remarkable collection of properties. It travels in a tightly collimated beam, is all of a single color, and is of great intensity. Their unusual properties make lasers extremely versatile, able to guide a missile to its designated target, perform precise alignments, and even serve as a pointer in the hands of a speaker presenting a visual display. Since their invention nearly forty years ago, lasers have proliferated into nearly all walks of life; they serve as the heart of systems for surgical cutting, communications, surveying, industrial welding, weapons guidance, displays, and a panoply of other applications.

An intense pulse of laser light, less than a billionth of a second in duration, leaves the telescope at the Observatoire du Cote d'Azur near Grasse, France, on its way to the moon and back. The pulse will be used to determine, by means of precise timing, the exact distance to the moon. Such modern refinements of the original 1969 experiment have allowed the determination of the moon's distance to within an inch.

The laser has helped usher in what is sometimes referred to as the "Age of Optics." A whole industry has grown up to manufacture cheap and reliable communications lasers used for fiber-optic transmission systems, and indeed the scenario of a laser in every home to handle the transmission of vastly increased amounts of data carrying communications, information, and entertainment is likely to become a reality within the next decade. And the development of robust cheap lasers has encouraged their use in countless new applications; examples include the laser readout head in an audio CD player or the rugged laser a plumber uses to align the drop in a drainage line.

Paradoxically, this most practical of devices could not have been created without the most cerebral and arcane development of twentieth-century physics, the theory of quantum mechanics. Early in the twentieth century a revolution in physics transformed our understanding of the structure of the atom. The best minds in physics grappled with an increasing body of experimental evidence that the classical description of the physical world was unable to explain. Newton and his successors had viewed the physical world as essentially deterministic: anyone with enough information should be able to predict the outcome of a physical process—the trajectory of a planet, the course of a chemical reaction. This concept of the world as being entirely predictable and deterministic had to yield to a newer understanding: at least at the microscopic world of the atom, physicists could predict only which outcome was most probable—there was always the chance that the most likely outcome would not be the one that was eventually observed. Einstein's oft quoted retort that "God does not play dice" proved surprisingly incorrect. And with this new understanding came the realization that characteristics such as the energy of electrons within an atom cannot, as in the classical view, take on any in a continuum of values, but instead could take on only values belonging to a limited, discrete set.

A laser is a truly quantum mechanical device in its operation since it makes explicit use of transitions between these discrete energy levels described so completely by quantum mechanics. Thus the quantum mechanically determined structure and properties of atoms are intrinsic to the workings of lasers. The ruby laser's light is nothing more than a collection of light particles emitted by the ruby's chromium atoms, and atoms indeed are the source of light in most of today's lasers. Thus an exploration of atomic structure and its effect on the atom's properties is at the core of our investigation of the laser.

The new science of quantum mechanics implies that it should be possible to create light with the special properties of laser light, but it was not for two or three decades after the theory had been worked out that someone understood how to do so. In the mid-1950s, Charles H. Townes, a professor at Columbia University, built a device that created a powerful guided beam of electromagnetic radiation, similar to light but of much longer wavelength. Townes' beam was invisible, and of limited usefulness, but scientists and engineers were tantalized by the practical potential of an analogous beam of visible light. When Townes soon after proposed a way in theory to repeat his success, but using visible light waves, his challenge was taken up by scientists employed in the emerging industrial laboratories of the postwar era. Scientists at companies like Hughes Aircraft Corporation and AT&T produced the earliest working lasers, first a ruby crystal laser in May 1960 and then a gas laser at the end of the same year. The interplay of basic science and industrial research has continued to guide the development of lasers in the years since.

These scientists and engineers saw immense potential in the collimated, uniform light of a laser. Let's return to the action in the moon-ranging experiment, since examined in greater detail it illustrates dramatically why, for many purposes, laser light is superior to ordinary light.

Laser Light versus Ordinary Light

After the laser at Lick Observatory has emitted its powerful beam of light, a series of prisms and mirrors diverts the ultrashort pulse into the main bore of the huge telescope. The lasing process selects out only those light beams able to reflect back and forth many, many times without drifting sideways out the crystal's side. This self-selection process results in a light column emerging from the telescope that diverges only a few ten thousandths of a degree from a perfectly straight line. The burst of light exits the top end of the telescope and travels upward into the night sky toward a precisely targeted area in the Sea of Tranquillity, on the distant moon, where only days before, on July 20, 1969, Apollo 11 astronauts Neil Armstrong and Buzz Aldrin have taken man's first steps on that surface. There, left behind by the astronauts, is a small 10×10 array of fused silica corner reflectors, similar in design to those employed as automobile taillights and reflectors on bicycles. The entire array, measuring a mere 18 inches across, faces directly back toward Earth.

Yet despite the beam's extreme collimation, by the time the pulse has traversed the intervening quarter of a million miles of space to reach the surface of the moon, its ever so slight divergence has expanded the beam to a diameter of about a mile. Hence most of the pulse spreads across the laser-illuminated square mile of the moon's surface without hitting the designated reflector array. If we were to break up the intense pulse initially emanating from the telescope on Earth into

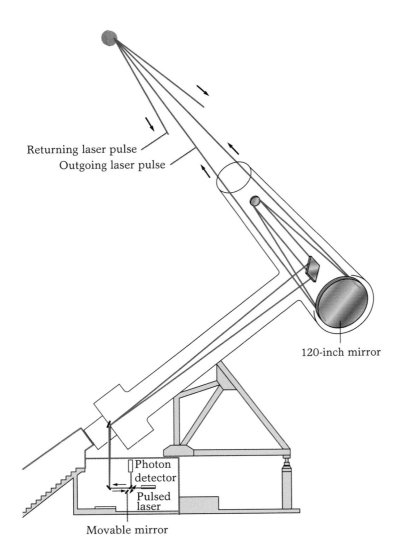

Returning laser pulse

Outgoing laser pulse

120-inch mirror

Photon detector

Pulsed laser

Movable mirror

A system of prisms and mirrors directs the laser pulse through the telescope at Lick Observatory on both the outgoing and incoming portions of its two-and-a-half-second round-trip to the moon. Shortly after the pulse leaves the laser, a lens expands the beam to just fill the aperture of the 120-inch telescope; sending the beam through the system "backwards" will still further decrease its divergence from its already low value when exiting the ruby crystal. On leaving the telescope the beam is 10 feet in diameter and is diverging by only about one foot in 50 miles. On the beam's return trip, a small mirror is flipped into position to deflect the beam into the detector.

its smallest fundamental physical units, known to physicists as "photons," we would be able to count a phenomenally high number of 100,000,000,000,000,000,000 (10^{20}) photons in the original laser pulse. Yet of that amazingly large number beginning the trip from Earth, only about 10 to 100 billion (10^{10} to 10^{11}) or so hit the reflector. Their encounter with the reflector array reverses their direction of travel, sending the photons back to Earth.

Back at the Lick Observatory, the telescope's optical system gathers the part of the beam that falls upon its open end. By this time, beam spreading and atmospheric absorption have whittled the beam down to

The Apollo astronauts, whose footprints remain frozen in time on the moon's surface, left this laser reflector there to make possible the laser-ranging experiment.

only a few photons. The signal is now almost infinitesimally small, and easily swamped by stray light from the moon's surface. To distinguish the exceedingly small signal from this potentially large background of unwanted "noise," the scientists make use of a crucial property of lasers: the light in the beam is emitted at a precise frequency determined by the internal structure of the chromium atoms, meaning that it is of one very well defined spectral color. As a result, the instrument at the bottom end of the telescope can be equipped with a filter that only allows the passage of photons of that precise color, virtually eliminating photons from other sources. The photons that make it through this screening device, now reduced to only a few in number, are directed into an optical detector known as a photomultiplier tube, whose sensitive electronic circuits can detect the arrival of even a single photon.

Despite the screening device, there is some slight chance that one or two detected photons are strays that did not originate from the initial laser pulse. So the scientists repeat the experiment again and again throughout the night, each time charging the laser with the energy to fire another intense pulse, and each time recording the arrival times of the possible returning photons. After a number of hours, a clear statistical pattern emerges, giving the precise arrival time of the returning photons and so the precise duration—slightly more than two and one half seconds—of the round-trip journey at the speed of light. From that well-known speed, the scientists at Lick Observatory are able to precisely convert the travel time into the round-trip distance the photon pulses have covered, allowing them to measure the approximately 240,000-mile distance to the moon to within a mere 15 meters.

In the more than twenty years since that night in 1969, scientists have refined this moon-ranging technique to achieve an ever increasing degree of precision. Harnessing the power of the laser, they have been able to determine the moon's distance to within an inch! Scientists have examined the variation in this distance over the course of many years to study phenomena as diverse as the wobble of the Earth's axis, the exceedingly slow drift of continents here on Earth, and even the predictions of general relativity, all unlocked by this powerful new tool we know as the laser.

Although scientists have learned to use light in a variety of ways to explore the world around them, the laser has allowed them unheard of precision. It would not have been possible to determine the distance to the moon using conventional light sources, which lack some of the key properties we have come to associate with lasers. Among them is the property of extreme collimation. Conventional light sources rely on a hot glowing object as their source of illumination. Whether the source is the white-hot tungsten filament in a standard light bulb or the electrically heated electrode tips of a spotlight's carbon arc lamp, it will by its very nature emit light equally in all directions. These conventional light sources can be made directional by partially surrounding them with a special curved reflecting surface, designed to bounce the light rays heading outward from the central point of the light source in such a way that each ray is directed along the same chosen direction, as shown in the left-hand panel of the figure on the following page. Such parabolic reflectors are employed in the directional spotlights that illuminate a theater stage as well as in the everyday household flashlight. It can be mathematically shown that, if the light source were truly a point lying at the perfect focal position indicated by the small yellow spot in the figure, all the light rays emanating from it would be directed along a single direction, and this perfectly collimated beam would rival and even surpass that of the laser.

The problem is that conventional sources of light, the glowing filaments and arcs, have some finite size to them. The more light that is required, the larger the source must be. Unfortunately, light that originates

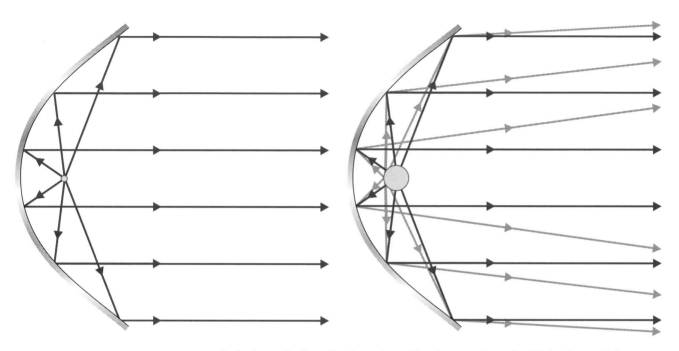

Left: A parabolic reflecting mirror like that used in a flashlight directs light rays from a conventional point light source, such as a hot tungsten filament in a light bulb, into a collimated beam. Right: Light rays emanating from the center of a larger source, indicated in dark orange, continue to be collimated by the reflector. In contrast, light rays from outlying points, such as those shown in light orange, tend to spread out, or "diverge."

from points away from the focal position no longer travels in the perfect collimated direction. The right-hand panel of the figure shows in light orange the paths followed by beams originating for example from the point closest to the mirror on the surface of a larger spherical source. You can see that light from this point fans outward. Light rays from other points on the glowing object's surface will also diverge, creating beams that are no longer parallel and that are misaligned with respect to the horizontal axis. For this reason, the brighter the spotlight, and hence the larger the glowing source, the less collimated the resulting beam.

We can make the beam more collimated by lowering the curvature of the reflector and moving it farther from the point source, thus making the source appear effectively smaller to the reflector. This procedure significantly reduces, but does not eliminate, stray noncollimated light. Unfortunately, it achieves collimation at the expense of efficiency since the more distant reflector traps fewer of the rays coming from the source. Headlights and lighthouse beams actually achieve better overall visibility from having some of their light projected in a slightly uncollimated way outside the main beam, so they employ highly curved mirrors tightly wrapped around the light source. Military signal lamps used for ship-to-ship signaling, on the other hand, are designed to minimize the stray light in order to avoid interception by enemy ships. Consequently, they employ less-curved reflectors positioned significantly away from the source but pay a penalty in lower beam intensity.

The main problem with conventional light sources is that there is no way to place more and more glowing material at a *single* point in space to make its intensity grow. The laser frees us from this constraint. Each atom in a laser's ruby crystal will emit a beam along the laser axis no matter where that atom is in the crystal, so we can put more chromium atoms into the ruby, or even make the crystal longer, to increase the beam's intensity. This is one of the reasons that lasers can be made so powerful.

The advantages of the laser do not end here. For example, an incandescent light or other conventional light source cannot emulate the laser's ability to release all the energy of its light beam in a few billionths of a second: the filament cannot be heated and cooled that rapidly. In the moon-ranging experiment, it is the pulse's ultrashort duration that allows its departure and arrival times to be determined so precisely.

A final distinguishing property of lasers is the fact that laser light consists solely of a single color. It is in optical parlance "monochromatic." The detector in the moon-ranging experiment made use of this intrinsic property of lasers when it filtered out virtually all the other light arriving within the view of the telescope. Here again, the laser has transformed the light from the flash lamp: it has converted the white light of the lamp into light of a more useful single color.

This intriguing light source has a number of other exotic properties that make it ideally suited to many novel applications. One, for example, is that the light waves in the beam undulate up and down in such a way that all the peaks and valleys of the distinct rays oscillate together in phase with one another, a property referred to as "coherence." Since conventional light sources are not coherent, new applications that make use of this property, including the creation of three-dimensional images by means of the technique called holography, had to await the invention of the laser.

In the years since the laser's invention, researchers hoping to exploit its properties in new ways have created radically new forms of the device. Perhaps the most important of the new lasers has been inspired by a revolution in the field of solid-state physics. This subdiscipline of physics came into its own after the elucidation of quantum mechanics opened a window onto the special properties of tightly packed atoms.

The highly curved mirror of this British reflector lamp, developed in the late-eighteenth and early-nineteenth centuries, directs the light beam at the lighthouse at Cape Bonavista in Newfoundland, Canada.

A series of ultra-high-power lasers converges on a small capsule of heavy hydrogen in a laser fusion experiment. The simultaneous arrival of the laser pulses from many different directions induces an implosion strong enough to set off thermonuclear fusion in the hydrogen, akin to the fusion reaction that serves as the energy source in the interior of the sun.

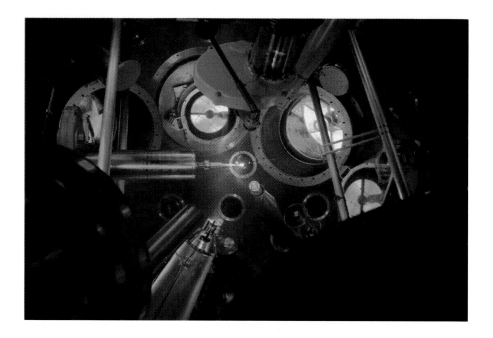

Scientists began exploring how individual atoms come to interact with one another as they are brought closer and closer into the close assemblage we refer to as a solid. In the course of their investigations, they discovered the potential of the semiconductor, which forms the basis of modern electronics. Semiconductors are materials that can change dramatically back and forth between electrically conducting and nonconducting states by the simple additions of carefully chosen impurity atoms and externally applied electric fields and currents, an ability that has made these materials the heart of today's "Information Revolution." In the 1960s, solid-state physics and lasers joined forces with the invention of the semiconductor laser, a class of laser very different from the ruby crystal laser. The semiconductor laser serves as the light source of today's globe-encircling optical communications network, carrying voice, data, and video signals across town or around the world at literally the speed of light. These devices rely not on individual atoms to create their magical light, but rather on properties of the overall semiconductor solids from which they are constructed. Our exploration of the laser will offer us the chance to examine the inner workings of solids from the standpoint of a solid-state physicist.

Semiconductor lasers are rich in variety and function, and to a large extent their myriad forms are made possible by advanced techniques for growing semiconductor crystals that have evolved side by side with new laser designs over the past three decades. So our journey will take us from solid-state physics into the realm of the materials scientist for a behind-the-scenes look at some of the laboratory techniques used to control the growth of semiconductor crystals. The most advanced techniques can literally grow a crystal atomic layer by atomic layer. They have made possible a laser so small that a million of them would fit on the nail of your little finger!

Despite the variety of lasers that exist today, we are only beginning to explore the possibilities. Scientists continue to push new frontiers in the development of these devices. They are extending the output of lasers to new spectral regions such as the X-ray and far infrared, and shortening pulse durations to less than 1/100,000,000,000,000 of a second (10 femtoseconds). In state-of-the-art research facilities, lasers are trapping beams of atoms to cool the trapped atoms to a temperature below a millionth of a degree above absolute zero, and they are imploding miniature capsules of heavy hydrogen to release the tremendous power of nuclear fusion.

As the Age of Optics continues to unfold, the fascination exerted by this exciting new light source has never been greater. As we explore the past and future of lasers, along the way we'll linger to discover what makes these marvelous devices work and explore in depth the remarkable scientific underpinnings in the fields of chemistry, physics, and materials science that have made lasers possible. Rarely has the interplay of science and technology offered a more intriguing story than the story of the laser.

The sources of light in this jungle of neon signs which makes up the glittering Las Vegas cityscape are transitions of electrons from one energy level to another within individual atoms. Such electronic transitions also serve as the fundamental light sources for lasers.

2

Light from the Atom

Imagine yourself standing in a meadow on a sunny summer day. The sun is shining brightly and the sky is blue. Trees stand in the distance, and wild flowers of yellow and blue are scattered through the lush grass. The grass is not monotone green in color, but as in an Impressionist painter's palette it appears to the eye as a host of different shades of green, with daubs of browns, yellows, and earth tones. The sunlight on your head and shoulders feels warm.

This scene, this seeing, this warmth, each depend on light. We can be aware of this scene without having to think about light, but we know intuitively that it is light from the sun that makes this scene, this awareness, possible. Light rays scattering from the objects around us into our eyes allow us to see and to function in life. At night we artificially generate light to illuminate our surroundings as best we can using incandescent bulbs, fluorescent tubes, even flashy neon signs. As we shall see, lasers are made from similar light generators, but the light is manipulated to form a highly coherent beam.

But what is light? The "stuff" of light remains an illusive quantity, even when we are told that it is, in physicists' terms, a form of electromagnetic energy. This is so, even though we might understand its energy character intuitively, having enjoyed the warmth of sunlight on our shoulders. Physicists further add to the mystery even while explaining that light has a dual nature, sometimes appearing as waves and sometimes as particles called photons. Just what, then, is the "stuff" of light?

Light As an Electromagnetic Wave

We can begin to appreciate the answer to the question "What is light?" if we look carefully at the regions in space near a magnet or charged object. Most of us have seen the remarkable effect a bar magnet has on a sprinkling of iron filings in its vicinity. The iron particles arrange themselves in arclike patterns, indicating the presence in the space near the magnet of something we call a magnetic field. Any magnetic or magnetizable material placed in the space containing this magnetic field is seen to react, being either attracted or repelled, as any child playing with a set of magnets discovers.

Something similar takes place in the space around an electric charge. Two charged particles placed near each other will react, showing either attraction or repulsion. The reader may have seen the classic presentation of this effect that is performed with two light Styrofoam balls hung by strings from the same point. Touching the balls with a rubber rod that has been rubbed vigorously with a piece of fur produces a dramatic response: the balls rise upward and apart from one another, seeming to defy the force of gravity. What has happened is that excess electric charge, in the form of electrons, has been rubbed off the fur and onto the rod, giving it a negative charge. When the charged rod is touched to the balls, some of the excess electrons, repelled by the others, move onto the Styrofoam balls. Since the two Styrofoam balls are both now charged with electricity of the same type, the negatively charged electrons, they repel each other, rising upward and separating.

A charged object, in some sense, creates a condition in space such that another charged object in this space is subjected to a force. It is this condition in space that scientists call an electric field. So, too, a magnetized

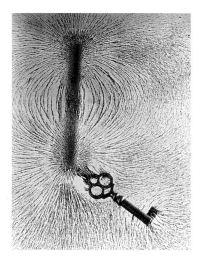

Iron filing particles in the vicinity of a magnet arrange themselves in patterns that reveal the geometry of the magnetic field. This field can be thought of as lines that emerge from one end of the magnet and loop around in space back to the opposite end. The key, containing iron, becomes magnetized by the magnetic field of the bar magnet and thus affects the field's general pattern.

object creates a magnetic field. The reach of such forces across intervening space is quite remarkable; but even more remarkable is what happens when the objects that generate these fields start to move.

The movement of electrons, or of any object carrying electric charge, by definition produces what we refer to as an electric current. Yet the purely electric phenomenon of an electric current creates, as a byproduct of its motion, a magnetic field. Thus when one passes a current through the wire windings of an electromagnet, a corresponding magnetic field appears around these windings. Similarly, a changing magnetic field can create electric fields. For example, a loop of wire spinning in the field of a fixed magnet experiences, as it turns, a changing magnetic field within its interior, and the result is an electric field along its length that drives electric current in the wire. This is precisely the physics behind how an electric generator works.

Working out the relationship between electric and magnetic fields, and how both fields change with time in response to one another, was the central effort of nineteenth-century physics. Its crowning achievement was the mathematical elucidation of these relationships in four relatively simple equations by the British physicist James Clerk Maxwell. Not only do these relationships, known as Maxwell's equations, predict the effects responsible for the electromagnet and the electric generator; they also provide the underpinnings for the phenomenon we know as light.

According to Maxwell's equations, an electric charge oscillating up and down produces an oscillating magnetic field, a simple extension of the electromagnet concept just discussed. But the changing magnetic field that results must in turn produce a changing electric field! The process is in a sense cyclical, and self-reinforcing. A careful examination of Maxwell's equations shows that our initial oscillating electric charge produces an oscillating wave moving away from it such as the one shown in the figure on the following page. The traveling wave consists of an electric field that oscillates up and down, between a positive and negative value, in the same direction the initial charge was oscillating, and a magnetic field that oscillates back and forth in a direction perpendicular to both the electric field and the direction of the wave's propagation, as shown in the figure. These waves, known appropriately as "electromagnetic waves," are analogous to those found on the surface of a pond, but there is one key distinction. Whereas the water waves require the surface of the water in order to propagate, electromagnetic waves propagate on their own without a propagation medium. In fact, they are capable of propagating light-years across the vast emptiness of outer space.

All waves can be classified according to their wavelength and frequency. Wavelength is the distance in space between adjacent crests in

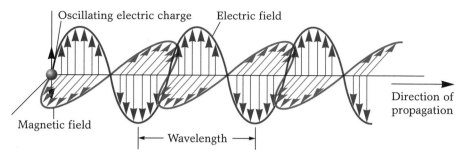

An electromagnetic wave, here shown emanating from an electron that is oscillating up and down, consists of oscillating electric and magnetic fields. The two fields are in phase and perpendicular to each other and both are transverse — that is, perpendicular — to the direction in which the wave propagates. Radio waves are examples of electromagnetic waves generated in this way by electrons oscillating up and down, typically in a metal antenna.

the wave, in this case corresponding to maxima in the electric (or magnetic) field strength. To interpret the frequency of a wave, imagine that you are focusing on a single point in space and observe the change in electric (or magnetic) field strength as the wave goes by. In the case of the pond, you would watch the water level move up and down at a single point in the water. The number of cycles from maximum to minimum to maximum strength occurring in a single second defines the frequency. Frequency units are thus cycles per second, now called hertz. Of course, there is a direct relationship between these two properties. The greater the wavelength, the longer we have to wait at a given point for successive maxima to arrive, and hence the lower the frequency of the wave. Expressed mathematically, the wavelength and the frequency of a wave are said to be inversely proportional to one another. Waves of longer wavelength have a lower frequency, while those of shorter wavelength have a higher frequency.

Electromagnetic waves exist in a wide range of wavelengths and frequencies. As shown in the figure on the facing page, the electromagnetic spectrum contains various kinds of waves including radio waves, microwaves such as those used in radar and microwave ovens, infrared-visible-ultraviolet light, X-rays, and gamma rays. All these waves are the same electromagnetic wave phenomenon; they differ one from another only in their frequency and wavelength.

Although the entire electromagnetic spectrum could be called light, this term is customarily reserved for those electromagnetic waves in the range around which the human eye is sensitive. The visible portion is relatively narrow, ranging from 700 nanometers (billionths of a meter)

THE ELECTROMAGNETIC SPECTRUM

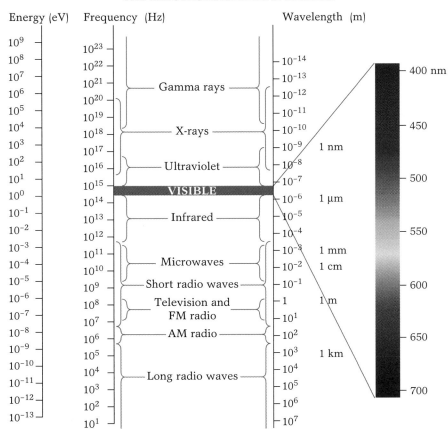

The electromagnetic spectrum spans a very large range from radio waves to gamma rays. All electromagnetic waves can be equivalently characterized by their energy, their frequency, or their wavelength in respective units of electron volts (eV), hertz (Hz), or meters (m), and all travel at the speed of light, 3×10^8 meters per second in free space. The visible light spectrum sensed by the human eye is just a tiny section of the overall range of electromagnetic radiation.

for deep red down to 400 nanometers for deep violet. Our eyes have evolved to be most sensitive to the color green, at wavelengths between 500 and 550 nanometers, near the middle of this visible portion of the light spectrum. In a visible light wave, the electric field, as well as the magnetic field, oscillates at a very high frequency as the wave propagates, cycling at a rate of 600 million million times per second (6×10^{14} hertz) for green light, for example.

Light waves, and in fact all electromagnetic waves, travel through space at a very high constant speed: 30,000,000,000 centimeters, or 186,000 miles, per second. These numbers are so big that perhaps a better way to look at this speed is to consider that light travels a distance of 1 foot (about 30 centimeters) in only one billionth (0.000,000,001) of a second (also called 1 nanosecond). In everyday life we seem to see events at the instant of their occurrence, but we know in our minds at least that this is not really true. Consider, for example, that the light from even the nearest star beyond the sun requires about four years to reach us.

Light: Particle or Wave?

When two waves cross each other in space, interesting effects occur at the places where they intersect. Unlike two colliding streams of water, for example, two intersecting light beams pass right through one another. Furthermore, in the region of space where they cross, we find that in some spots the light intensity is practically zero, while in other spots the intensity is larger than the intensity of either wave separately.

A simple law, known as the principle of superposition, explains these effects: two crossing waves sum at the point of intersection; the amplitudes of the resultant wave's electric and magnetic fields at that particular point are merely the sum of the corresponding fields of the two waves taken separately. For clarity, we concentrate here on the electric field, although the same effects hold true for the magnetic field. If one wave has a positive electric field at one point while the other wave has a negative electric field at that same point, they would add up to a value near zero and virtually cancel each other, an effect called destruc-

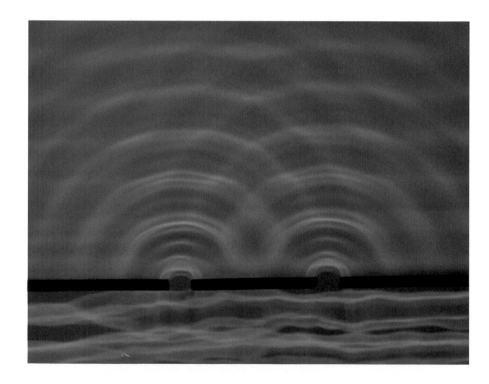

Water waves illustrate the effects of interference. The waves, entering from the bottom in the photo, pass through two openings in a horizontal barrier to create a pair of concentric wave sources in the upper half of the figure. The two combine through the principle of superposition to form a pattern representing points at which the two waves add and cancel respectively.

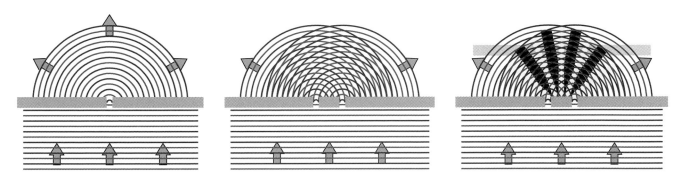

Light, consisting of a series of peaks (red) and troughs (blue), traveling through two narrow slits demonstrates wave interference. When light passes through a single slit, the parallel plane wave effectively becomes a narrow circularly radiating point source (left panel), but the same wave produces a more complex diffraction pattern on passing through a closely spaced pair of slits (middle and right panels). The yellow bands in the right-hand panel show the places where peaks add to peaks and troughs to troughs to form a brighter light intensity, and the dark bands show places where troughs and peaks cancel one another to form a null.

tive interference. At another point the waves might both have positive electric fields; they would add up to a value greater than either wave taken separately, an effect called constructive interference. Thus the region where the two beams cross shows a pattern of bright and dark regions, well explained by this wave model of light.

Such wavelike phenomena can give rise to unusual effects indeed. Consider the way a broad beam of light passes through a narrow vertical slit, shown from above schematically in the left-hand panel of the figure on this page. When the wave encounters the brown barrier, the barrier absorbs all but the small amount impinging on the narrow slit. The oscillating electric field that passes through this single slit fans out on the other side to become a radially propagating wave, akin to the pattern found on the surface of a pond on which a single source is bobbing up and down. In fact, this bending, or "diffraction," is precisely what happens when long straight lines of ocean waves impinge on a narrow breakwater opening to a harbor area.

The presence of diffraction is a clear indication that we are dealing with a wavelike phenomenon. If the light were merely a stream of particles hitting the barrier, they would form a narrow beam on the other side of the opening, rather than fanning out. The wavelike nature of light becomes even more evident when we consider what happens when the incoming beam encounters two slits, as shown in the middle panel of the figure. Now the waves radiate outward from each slit, and the fields of the two overlap with one another. At points where the peak (red curve) from one slit crosses the valley (blue curve) from the other, the waves cancel each other in destructive interference. These points are indicated by the broad dark lines in the right-hand panel, which connect all the points of crossing of the red and blue curves. At points of constructive interference, traced with broad yellow bands, the fields radiating from the two slits add together, and the light is most intense. The result is a spatially alternating pattern of bright and dark. If we were to insert a screen to observe these alternating light patterns, indicated by the tan line in the last panel of the figure, we would find a series of alternating vertical slices of light and dark, instead of just two bright vertical lines the size of our initial slits as we would expect if the light were simple beams of particles. Furthermore, as predicted by the right panel of the figure, the screen would be brightest right in the middle where the yellow line intersects with it (precisely where the particle model would predict it to be dark), directly between the two slits, and dark on either side (where the particle model would predict it to be lightest). This behavior—observed when the slits are made very narrow, on the order of the wavelength of light—confirms the wavelike nature of light.

But light sometimes has properties that a wave model cannot explain. Consider the so-called photoelectric effect. To observe the effect, light is shined onto a metal surface. The incoming light provides the energy needed to dislodge electrons from the surface; they are collected in the form of an electric current by a nearby electrode and measured. The wave model of light would predict that the more intense the incoming light, the higher the amplitude of the electric field in the incoming light wave, and hence the greater the number of electrons dislodged. In fact, this is what is observed experimentally, at least for certain wavelengths.

The puzzling effect is that above a certain wavelength, or alternatively below a certain corresponding frequency, absolutely no electrons are emitted no matter how intense the incoming light is. How can simply changing the spacing or frequency of wavefronts cause such an abrupt change? The wave model of light is insufficient to explain the phenomenon. An altogether different model is required.

This alternative model of light, first glimpsed in 1900 by the German physicist Max Planck, a pioneer in the creation of the new field of quantum mechanics, maintains that light consists of a stream of discrete packets of energy, or "quanta." Einstein was the first to propose light quanta and to use the concept to successfully describe the photoelectric effect in 1905. According to this alternative model, light can be thought of as a stream of particles, called photons; each type of light corresponds to a stream of photons of a particular energy. More precisely, the greater the frequency of the light, the greater the energy of the photons. This energy is usually expressed by physicists in units called electron volts (eV), where 1 eV is defined as the energy an electron gains when given additional speed under the influence of an electric potential of 1 volt over a distance of 1 meter. There is an exceedingly small amount of energy in a single photon packet, but since a light wave consists of an exceedingly large number of photons, the overall transport of energy by light can be significant, as indicated by that warm shoulder in the bright sunlight. The energy of a photon of green light lies between 2.25 and 2.48 eV, while that of an X-ray lies at values greater than 100 eV. The left-hand column of the figure on page 17 shows the photon energies in eV for photons across the entire electromagnetic spectrum. The more intense the light beam, the greater the number of these photon particles in it.

The photoelectric effect can then be understood in the framework of the photon model of light. Each electron requires a certain minimum amount of energy to be dislodged from the metal surface. Only if the energy of an incoming photon is higher than that minimum energy will it be able to dislodge the electron, forming in turn a stream of emitted electrons at the detector. Once the photons are of that sufficiently high energy, then the number of electrons emitted is directly related to the number of such photons incident. For light of longer wavelengths, or alternatively of lower energy per photon, each photon in the beam lacks sufficient energy to dislodge an electron, and hence turning up the brightness only increases the number of these ineffective photons impinging on the surface. No electrons are emitted.

It becomes clear when looking at a phenomenon like the photoelectric effect that light indeed sometimes behaves as particles. Physicists feel quite comfortable using both the particle and wave descriptions of light, and freely employ whichever one describes the phenomenon at hand the best. In fact, the quantum mechanical view of the physical world insists that the wavelike and particle-like properties are actually different manifestations of the same underlying structure. As we will see, not only can light *waves* be seen to possess particle-like properties, but *particles,* such as electrons and protons, can be seen to possess wave-like properties.

Before we turn to the question of where light comes from, an interesting related question is where all the light around us goes to. Light's fate is sometimes a source of confusion for people who remember enough high school chemistry and physics to recall that energy itself is always conserved, and never created or destroyed. Photons are just packets of energy in the form of light, yet we know intuitively that they can be created—by a light bulb, the sun, or a glowing fire. Yet the creation of photons is not in fact the creation of new energy, but the conversion of energy from one form—the electricity driving the light bulb, the fusion reactions within the sun, or the chemical burning of wood—into a new form, that of light.

In a similar way photons can also be destroyed. When we turn the lights off in the room, the light dies away. Where do the photons go? At first, they are reflected by the objects in the room. Photons of all colors fall upon the objects, but only photons of certain colors are reflected back, giving those objects their color to our eyes. The rest are absorbed. Eventually they are all absorbed by one object or another, and their energy is converted into a different form, most often heat. Thus after sunset light disappears and darkness falls, yet on a hot summer night we can still feel the warmth of the energy brought by the previous day's sunlight.

Atoms and Orbitals

At a superficial level the source of light is obvious: it comes from the sun, a burning fire, a light bulb, or even a laser. We shall focus here on the physical process that produces this light, for we will find in subsequent chapters that a laser is able to harness this process in an extremely elegant and efficient manner.

To discover the underlying process by which all light is generated, we need to journey into the microscopic world of the atom. There things are so small that they are measured not in units of inches or millimeters, but in angstroms, the unit defined as 10^{-10} meter! An angstrom is the width of your fist shrunk a thousand times, to roughly the thickness of a human hair, and a thousand times again, to the approximate size of a microscopic virus or about a fifth the wavelength of green light, and finally yet another thousand times. This tiny unit of measurement, a single angstrom, is the size of a small atom. It is in this submicroscopic world, just barely at the limits of resolution of the world's most powerful microscopes, that the electrons, protons, and neutrons of which all matter is composed interact with one another to form the atom. The particles in this tiny atomic world behave very differently from what we have come to expect of objects at our macroscopic scale. Actions are dominated by the physical principles of what scientists call quantum mechanics, the twentieth century's great contribution to modern physics.

At the center of the atom lies a tightly bound cluster of particles of two types: protons, positively charged subatomic particles, and neutrons, similar in size to the protons but possessing no net electric charge. This cluster, referred to as the atom's nucleus, is 10,000 times smaller than the ultrafine angstrom unit discussed above. If this angstrom-size atom were magnified to the size of Yankee Stadium, the nucleus would be the size of

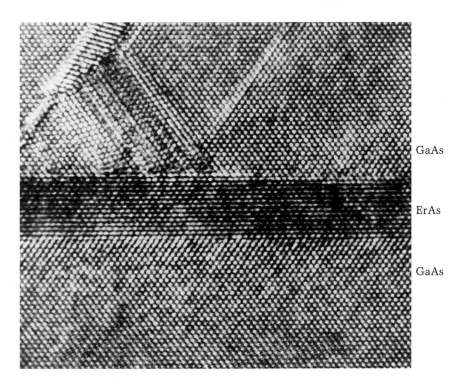

GaAs

ErAs

GaAs

An electron microscope image of a layered crystal structure in cross section. The magnification is high enough to just resolve individual atoms. The image was produced by transmitting electrons through a thin section of material that is about 100 atoms in thickness. Thus each white dot corresponds to a column of about 100 individual atoms lined up one behind the other in a well-ordered crystal structure. Three distinct layers can be distinguished, composed of gallium or erbium arsenide (GaAs or ErAs). The predominant diagonal decorations in the top layer indicate faults in the regular stacking of atoms, that is, defects in the perfection of the regular crystal structure. The scale is such that, in the top and bottom layers, the distance between two white dots in the vertical direction corresponds to 5.6 angstroms, equivalent to 0.56 nanometers.

a mere baseball. Yet this nucleus contains more than 99.9 percent of the atom's mass! The rest of the atom is essentially empty space . . . except for the surrounding swarm of negatively charged particles, the electrons, each of which is even smaller than the nucleus itself. But small and insignificant as they may appear, it is these electrons buzzing around the central nucleus that are the underlying source of almost all light.

The electrons in this ever-whirling world are often described as traveling in "orbits," although these orbits are far more complex than the simple elliptical orbits of, say, the planets around the sun. Electronic orbits do not lie in a single plane like planetary orbits; rather, they are fully three dimensional, and according to quantum mechanics are best described not by a single determinate path, but rather by a probability distribution indicating the likelihood of finding the electron in a given direction at a given distance from the nucleus. The use of these descriptions is analogous to describing a swarm of bees not by specifying the position and trajectory of each angry bee, but by giving the probability of being stung as a function of your relative position from the central beehive. Although these probability distributions are formally described by mathematical equations, in order to better visualize them they are frequently displayed graphically as a cloud surrounding the central nucleus. The cloud shading becomes darker in those areas in which the electron is more likely to be found at any given instant.

On the facing page you can see a series of such electronic orbitals, each with its own unique distribution. In the figure they are shown in rows of numbered "shells" of increasing average distance between the negatively charged electron and the central positively charged nucleus moving from the bottom row to the top row of the figure. Since opposite charges attract, it takes increasing amounts of energy to further separate the two charges, and hence the orbitals in the top row have the highest amount of energy associated with them, and the bottom row the least. In the so-called s levels in the left-hand column of the figure, the electron is distributed evenly in all directions, but the p and d orbitals have more complex shapes with various lobes extending out in preferred directions. It is these lobes, acting as outstretched arms, that interlock to form chemical bonds between atoms; these vital linkages are the basic building blocks of molecules and entire solids.

Although we can imagine an infinite variety of such orbitals, each describing a different distribution of electrons, in nature we find only a certain discrete set of orbitals, including those shown in the figure. What explains the existence of this distinct set? The answer comes again from quantum mechanics, and it requires looking at these tiny particles called electrons in the same way we have just looked at light. Like light, electrons can be described as possessing both wavelike and particle-like properties. When one describes the charge of an electron, one treats the electron as a discrete particle. Each such particle carries precisely one negative charge on it. And electric current consists of a stream of such particles flowing through an electrical conductor. This description seems to provide a nice complete picture. And it did . . . until experiments performed early in the twentieth century began to show the inadequacy of treating the electron simply as a particle.

Classic experiments performed in the 1920s by Clinton Davisson and Lester Germer at Bell Telephone Laboratories revealed something quite unexpected. We saw earlier in the chapter that if light falls upon a pair of closely spaced slits, it forms a pattern on the other side indicative of

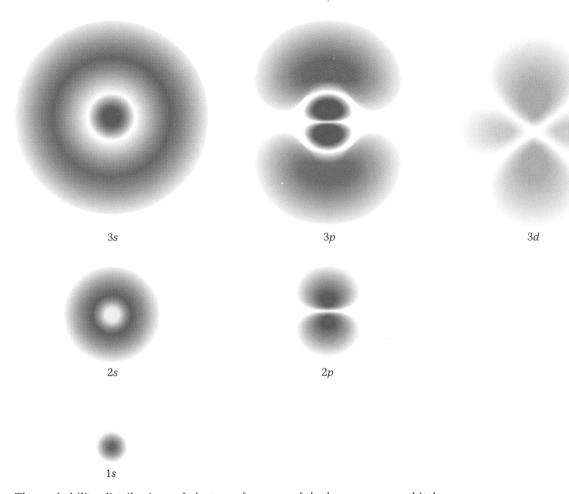

The probability distributions of electrons for some of the lower-energy orbitals in an atom, all drawn to scale. The full three-dimensional shapes of the orbitals are obtained by rotating the shapes shown here about the vertical axis. The probability of finding an electron is higher in more darkly shaded areas within the orbitals. The *p* orbitals shown are one of three possible orbitals (the other two point along the two other principal axes), while the *d* orbital is just one of five such orbitals. The average spacing of the electron from the central nucleus is roughly the same for all the orbitals in a shell (1, 2, and 3 for the rows from bottom to top).

interference. Such an interference phenomenon is inexplicable if we think of light as a continuous stream of discrete particles, each of which will intuitively end up going through one of the two slits. Instead we need to think of the light as a wave that passes through *both* slits simultaneously. The pattern formed on the other side is then completely describable mathematically as the interaction (or more formally the

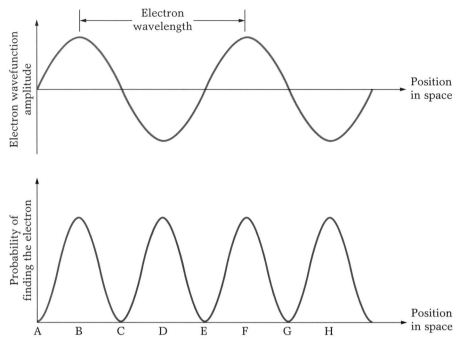

A representation of a simple wavefunction of an electron (upper graph). If such an electron were accelerated inside a vacuum tube through an electrical voltage drop of 50 volts, it would attain an energy equal to 50 eV, corresponding to a wavelength of 1.7 angstroms. Higher-energy electrons have correspondingly smaller wavelengths, so that an electron with an energy of 500,000 eV has a wavelength of 0.017 angstrom. When the electron wavefunction above is multiplied by itself (lower graph), the result represents the probability of finding the electron at various points in space along a given line. For this case, the electron has the highest probability of being found at points labeled B, D, F, or H. The electron has zero probability, and therefore will never be found, at points A, C, E, or G. Obviously the motion of this electron, as with any electron, does not correspond to the phenomena we are used to for large, macroscopic objects in our everyday world.

"interference") between the parts of the wave coming through each of the two slits.

Davisson and Germer observed how a beam of electrons passes through a thin metal foil. They discovered that a well-collimated beam impinging on one side of the foil produced on the other side not a single outgoing beam, or even a simple pattern explainable as individual electron particles scattering off atoms in the metal one at a time, but a complex pattern resembling an interference pattern. After months of careful consideration, they concluded that such a structure could only be the interference pattern resulting from the simultaneous scattering of the same electron off many atoms within the metal foil. The electron behavior was truly a wavelike interference phenomenon, completely analogous to the behavior of light in the two-slit light-scattering experiment if only there had been many more slits.

The truly remarkable finding was not that the metal atoms scatter the electrons, which was well known at the time, but that the electrons being scattered must possess wavelike properties, with troughs and peaks similar to those in light waves. Like photons, electrons and other small particles are both particle-like and wavelike at the same time. And it is their wavelike properties that serve as a basis for understanding the discrete quantum mechanical orbitals of the atom's electrons.

The way in which quantum mechanics dictates the existence of only a limited number of discrete orbitals was originally elucidated by the famous Danish physicist Niels Bohr. Soon after, the French physicist Louis DeBroglie first proposed that electrons, like light, have both particle and wave aspects. Subsequently, physicists hypothesized that if electrons are wavelike, then they can be described by a wavelike function. This wavefunction—which oscillates up and down, having both positive and negative values—does not literally represent the path followed by the electron. Rather, in the quantum mechanical model, the value of the wavefunction multiplied times itself at any point in space—which is always a nonnegative number—represents the probability of finding the electron at that point. Waves always repeat themselves, and this is of course true of the electron wavefunction as well. An electron moving uniformly in a straight line has the simple oscillating wavefunction shown at the top of the figure on the facing page. Points on the horizontal axis represent the position in space, and points on the vertical axis give the value, or "amplitude," of the wavefunction at each position. It is the square of this amplitude, plotted in the bottom part of the figure, that has the important physical significance of representing the probability of finding the electron at a point in space.

If one starts at a point within the atom and, instead of traveling along a straight line in space, traces for simplicity a circular closed path of a particular electron traveling around the nucleus of the atom, this undulating quantum mechanical probability wavefunction will alternate up and down along this closed orbital path as well. The fundamental insight was to realize that once the path of the orbit returns to any arbitrarily chosen starting point, the value of this quantum mechanical probability must of necessity have undulated back to precisely the value it had at the beginning, or else the electron would interfere with *itself* in its own orbit. To put it another way, the only stable orbits (i.e., the only ones that do not

One-integral-wavelength orbital

Two-integral-wavelength orbital

Three-integral-wavelength orbital

In the simple Bohr model of the atom, the electron orbits of an atom are stable only when the circumference is precisely an integral number of electron wavelengths, each indicated here by a complete cycle through both positive values (orange) and negative values (blue). For the three orbits shown here, the circumferences are 1, 2, or 3 electron wavelengths.

destructively interfere with themselves) are those described by an undulating electron probability that alternates up and down an exact integral number of times during the path of one complete orbit.

This Bohr model of the atom asserts, therefore, that only certain atomic orbitals are quantum mechanically "allowed." They are in fact more complex than the simple circular loops depicted on this page since they are truly three dimensional, like the electron probability distributions depicted on page 25. But these more complex orbitals still rely on the simple concept that only integral shifts in the electron's wavefunction are allowed in complete circuits around the nucleus. Bohr's finding is quite profound. As mentioned above, each of the possible, or "allowed," orbitals has a different specific spatial separation between the negatively charged electron and the positively charged nucleus, and it takes different amounts of energy to separate electric charge to different distances. Therefore each orbital has its own unique energy. Since there are only certain allowed orbitals, this means that there are only certain allowed energy states, or levels, that the electron can assume. Although the concept of discrete energy levels as distinct from a continuous spectrum of allowed energies may seem counterintuitive, it is one of the most important outcomes of a quantum mechanical treatment of the atom. And it is at the heart of the answer to the question we began this discussion with: Where does light come from?

Light Absorption

Before we return to the question of how an atom emits light, we need to understand how an atom *absorbs* light, since the absorption of a photon of light and the emission of a photon are precise reverse analogs of each other. The absorption of light by an atom is intimately tied to the "discreteness" of allowed energy levels just discussed, and it relies on an underlying property of electrons named the "Pauli exclusion principle" after the Austrian-born physicist Wolfgang Pauli. The principle states that no two electrons with the same value of "spin" can occupy the same orbital (or as Pauli would say, "the same quantum state"). Since the spin value of an electron can take on only one of two values, "up" or "down," the Pauli exclusion principle implies that there is room for just two electrons in each distinct orbital. Just like territorial animals, particles such as electrons each require their own "space," and each electron in an atom has its own unique combination of spin and orbital. This simple fact has profound consequences for the structure of the atom.

For an electron orbiting around the central nucleus, there are then a whole series of possible orbitals, as first proposed by Bohr, each orbital with a specific energy. An analogy might be the seating sections in a broad amphitheater, sloping upward as one moves away from the central stage. Each section represents an orbital, a "position" in space where the atergoers may sit. The quantity analogous to the atomic orbital's energy might be its distance from the stage—how far you have to climb to your seat. Applied to this analogy, the Pauli exclusion principle would assert that, as the theater fills, each seat can be occupied by no more than one individual at a given time. As the first theatergoer enters the amphitheater through a door next to the stage, he or she will choose the section closest to the stage—in a seat with the lowest "energy" in our analogy. Analogously, the first electron added to our atomic nucleus in building up our atom will, in trying to expend the least amount of energy, settle into the lowest energy orbital of the atom. New audience members arrive and take seats in this closest section until it is completely full. Thereafter, when new people arrive, they must be seated in the next section back. In such a way, the theater fills from the bottom up.

In the atom the orbitals are analogous to the seating sections. Each orbital holds just two electrons, two "seats" per section, one for "spin up" and one for "spin down." Just as our theater may have groups of sections that are roughly the same distance from the stage, such as the front left and the front right, so too our atom can have orbitals with roughly the same energy, such as the $2s$ and the three $2p$ orbitals depicted on page 25. A group of orbitals with roughly the same energy is referred to as a shell.

So our atom fills with electrons from the bottom shell upward, filling the slightly lower energy orbitals within a shell grouping first, and placing precisely two electrons in each orbital filled. Thus a *complete* atom, in its lowest energy configuration, consists of a number of "filled" lower-energy states, together with a series of higher-lying "empty" or "unfilled" states, analogous to the filled close seats and the unfilled seats farther up in our partially filled amphitheater.

If our audience members enter the theater with a bit more energy available to climb, then some may prefer to seat themselves farther up in the theater, to get a more panoramic view or perhaps to put a little more distance between themselves and their immediate neighbors. This in turn will result in some of the closer seats that would have been filled in our first scenario being left empty.

Something analogous can happen to the electrons near the top of the filled energy states, or orbitals, in our atom. If given the proper amount of excitation, they can be promoted into a higher unfilled state. The necessary energy is precisely equal to the difference between the energy level of the initial filled state that the electron is leaving and the energy level of the initially empty state into which it is being excited, the extra distance from the stage a misanthropic audience member chooses to climb when abandoning his or her close seat in favor of one higher up in our analogy. Although the list of all possible distances from the stage is well determined for a given theater, that list is in principle completely different in its details for a different theater with a different seating arrangement. So, too, the set of discrete energy levels that can be potentially occupied in a given atom is in principle different in all of its particulars from the discrete energy levels of a different kind of atom. For each different combination of protons and neutrons that determines a particular type of atom, there is therefore a unique set of energy levels. In fact, these differences form a unique fingerprint for each type of atom.

We can now see what all these energy levels have to do with the absorption of light. There are a number of possible ways for an electron in an atom to be elevated to a higher energy level, but one way is by absorbing a photon of light of energy just equal to that extra energy elevation. A remarkable side effect of this behavior is the fact that an atom cannot absorb light of any arbitrary energy, but must instead absorb energy only in discrete amounts that depend on *that particular atom's* allowed energy-level structure.

The discrete nature of the atom's absorption of light is often used by scientists to identify unknown gases, which after all are just large groups of isolated individual atoms like the ones we have just discussed. The identification is performed by passing a broad continuous spectrum of light, containing all possible colors and hence all energies in the spectrum, through the gas and analyzing what colors are allowed to pass through. One finds a series of "holes," or dark lines, in the spectrum of light that has passed through the gas. Each hole corresponds to a difference in energy between one of the filled energy states and one of the empty states in the atom. Atoms in the gas readily absorb photons of that energy, leaving the hole in the spectrum, and the energy of each absorbed photon is used to promote an electron to a higher energy state. The energies of the dark lines in the spectrum on the following page correspond precisely to the vertical arrows in the adjacent panel, which indicate just such allowed transitions. Since, as we mentioned earlier, each atomic element has a different set of such levels, this series of absorption lines serves as a unique fingerprint for a given chemical element in the gas. In fact, this observation that atoms absorb only at such discrete energy levels, unexplainable by the deterministic world of Newtonian physics, was one of the puzzling phenomena that stimulated the development of the new science of quantum mechanics.

In the case of atoms in a gas, each atom acts as an isolated system, independent of all the atoms around it, since they are all exceedingly far away from one another on the scale of the dimensions of an individual atom. As such, the electrons associated with each atom have orbitals localized to that particular atom, with the discrete energy levels de-

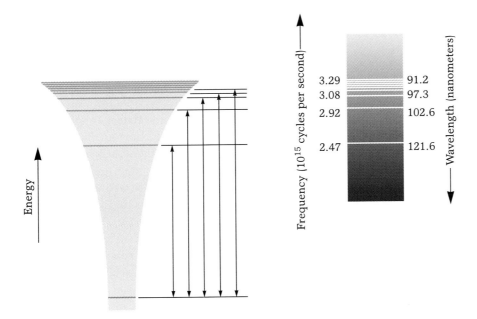

Electrons jumping to higher energy levels in the hydrogen atom (left) produce the absorption spectrum at the right. This spectrum, the so-called Lyman series of hydrogen atoms, represents what remains of a continuous spectrum of light once it has passed through hydrogen. It contains missing holes, or what are referred to as "dark lines" since there is no light present at these particular lines in the spectrum. They occur at specific wavelengths, or specific frequencies, in the ultraviolet region of the spectrum. For example, an electron in the lowest energy level can jump to the next higher level by absorbing a photon of wavelength 121.6 nanometers, resulting in a dark line in the spectrum at this exact wavelength.

scribed above. In the case of a solid, however, the atoms are much more tightly packed, and the spacing between atoms making up the solid becomes comparable to those very same atomic dimensions. In this case the system becomes a great deal more complex. Only the outermost electrons, the ones in orbits farthest from the nucleus, are significantly affected.

Suppose one assigns each of the filled electron orbitals a number that corresponds to its average distance from the nucleus; one can then order the various orbitals into "shells" of increasing distance as discussed earlier. In order for the atom to be electrically neutral, it must have a number of positively charged protons in its nucleus equal to the number of electrons in the filled shells surrounding it. The electrons occupying the innermost shell feel the attractive force of that entire positive charge and so become tightly bound to the nucleus. As we move outward in shells, the electrons continue to feel the influence of that same positive

Absorption Spectroscopy and Astronomy

The concept of a unique fingerprint in the absorption spectra of each element serves as a powerful tool in astronomy and astrophysics. In fact, the only way astronomers have of determining the chemical nature of the universe beyond our own Earth and the handful of nearby planets to which we have been able to send spacecraft is to look for these characteristic absorption lines. By pointing a telescope at a single luminous object in the sky such as a star or galaxy, the astronomer can analyze the "dark-line" spectrum of the arriving light. By matching that spectrum with known atomic spectra obtained in a laboratory here on Earth, he or she can determine the identity of the absorbing chemical species. In the case of a star, these represent the atoms in the outermost layers through which the light generated in the inner regions of the star must pass on its way out of the star en route to the astronomer here on Earth. This is precisely how the element helium was first discovered in the sun before it was found on Earth. Such observations can even detect the presence of "dark matter," interstellar material such as clouds of gases that are not hot enough to glow and hence cannot be directly detected with an optical telescope. Astronomers analyze the characteristic missing lines in the light from a distant star after that light has traveled through vast expanses of this intervening dark matter, and by this means they can conclusively determine the elements that make up this "invisible" material.

Spectral shifts of these "fingerprints" toward either the blue or red end of the spectrum indicate relative motion of the absorbing object either toward or away from us, in the same way that a train whistle changes pitch, or frequency, as the train passes. The pitch is higher as the train approaches, and lower as the train moves away from us, known as the Doppler effect. Red shifts in frequency are measured for all distant galaxies, indicating that they are receding from us, a key experimental clue that led eventually to the "Big Bang" theory of the creation of the universe.

attractive nucleus within, but this attraction is partially counteracted, or "screened," by the repulsive effect of the electrons in filled shells closer in to the nucleus. Thus electrons are more and more weakly bound to the nucleus as one moves out in shells. It is the most weakly bound electrons, those in the outermost shells, that participate predominantly in the formation of the solid.

As the atoms forming a solid assemble, each individual atom brings along a few orbitals filled with weakly bound electrons, together with a handful of similar higher levels that are not occupied by electrons. The outer reaches of these levels begin to touch one another as the atoms get closer to one another within the solid. An electron just halfway between the two atoms feels itself as part of orbitals from *both* the neighboring atoms. We say the electron clouds begin to overlap. The individual atomic orbitals lose their exclusive connection to a single nucleus and begin to change their shape and bridge from atom to atom, ultimately extending throughout the solid. Since a piece of solid you can hold in the palm of your hand contains on the order of 10^{24} atoms, each bringing a few orbitals to this merging process, we go from a few times 10^{24} individual orbitals, each on a given atom, to that same large number of orbitals all spread out across the whole assemblage of atoms. The result is a solid with a very dense number of orbitals, or "states" in the parlance of solid-state physics, on the order of a few times 10^{24}, each extending spatially throughout the entire solid. There are so many of these energy levels that they can no longer be described as discrete; instead they form a virtual continuum of energy states. At energies for which there are large numbers of levels from the initial atoms that have now merged, the density of levels, what the solid-state physicist refers to as the "density of states," is quite high. On the other hand, at energies for which our initial isolated atoms had no corresponding levels, there are no allowed states at all in our new solid, and the density of states falls there to zero.

So the picture of energy states has now changed in a solid. Instead of a handful of isolated discrete states associated with a single atom, in a solid we have a continuum of states as a function of energy (at least for those filled by the outermost higher energy electrons), and the density of those states varies from exceedingly high values at some energies to none at all at others. The result is the formation of broad bands of allowed energies, separated in energy by "forbidden gaps" that contain no allowed states whatsoever. This model of a solid is compared to that of an atom in the figure on the top of the next page. The filled levels which are so tightly bound to the nucleus that they do not overlap with adjacent atoms in the solid, referred to as "core levels," remain essentially unchanged in the transition. But the outermost filled levels of the atom merge to form filled energy bands (red bands), while the outermost empty allowed electron levels merge into empty energy bands (blue bands). The bands are separated by forbidden energy gaps, shown in yellow.

The figure on the bottom of the next page shows a schematic representation of the energy levels of one type of solid. We consider only two bands, the lowest empty blue band and the highest full red one. Since the core states play little part in the properties of the solid, they are omitted for the sake of simplicity. Just as the lower outermost levels in the isolated atom in its "ground" state are filled and the higher-lying levels empty, so the lower bands of the solid in its "ground" state are filled, as indicated by red, and the upper bands empty, as indicated by blue. In

The electron energy levels of isolated atoms remain discrete, whereas in a solid the discrete energy levels of the individual atoms now become bands of states that extend throughout the solid, although electrons in the inner orbits do not sense neighboring atoms and remain isolated. The forbidden energy gaps between bands, which have no states available for electrons, are derived from the equivalent spaces between orbitals of isolated atoms. Here the horizontal width of each energy level indicates the size of that orbital. The most tightly bound electrons, near the nucleus, have the lowest energy and the smallest orbital diameter.

An electron can jump from the lower filled energy band (orange) into an empty state in the upper empty allowed energy band (blue) by absorbing a photon of the energy required for the jump. Transitions either to or from states in the forbidden energy gap are forbidden, as indicated by the dashed arrows. Since electrons cannot jump into the forbidden gap, the solid is transparent for photons of energies less than that corresponding to the width of the forbidden gap. The semiconductor silicon, from which the integrated circuits of computers are made, is completely opaque to visible light but is transparent in the infrared region of the spectrum for photon energies less than 1.1 eV, corresponding to arrows shorter than its forbidden energy gap.

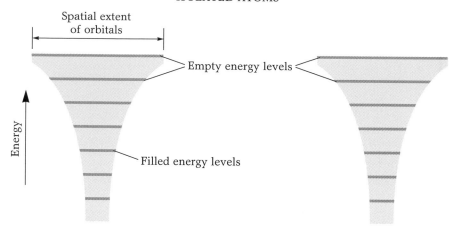

ISOLATED ATOMS

Spatial extent of orbitals

Empty energy levels

Energy

Filled energy levels

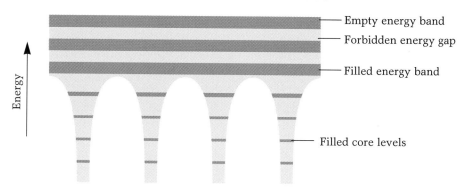

A SOLID OF OVERLAPPING ATOMS

Empty energy band
Forbidden energy gap
Filled energy band

Energy

Filled core levels

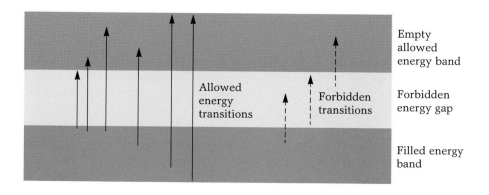

Empty allowed energy band

Forbidden energy gap

Filled energy band

Allowed energy transitions

Forbidden transitions

Transparency

The concept of transparency underlies the now classic balloon within a balloon demonstration used ever since the days of the first lasers to illustrate the power of these devices. A black balloon, made of dark, absorbing rubber, is inflated inside a larger clear balloon, made of relatively transparent rubber. In the demonstration, a bright laser beam is directed toward the pair of balloons. The central black balloon has many allowed atomic transitions (red to blue in our discussion below), and hence it is able to absorb many if not all of the large number of photons reaching it from the laser. As a consequence, it heats rapidly at the point where the laser beam hits. Quite quickly a hole burns at this point and the balloon bursts. Yet quite astonishingly to the casual observer, the outer balloon remains intact, unpopped! The laser beam has no effect on the outer balloon since its atomic structure does not allow for the absorption of the light in the beam.

It is precisely this phenomenon that is responsible for the success of laser eye surgery targeted at reattaching a torn retina. Because the atomic structure of the clear lens of the eye does not contain any filled-to-unfilled electronic transitions in the energy range of the visible laser

Visible laser light can pass through the transparent lens of the eye and deliver its energy to the back of the eye, where its absorption, and resultant heating effects, in the retina can be employed for reattachment in a "welding" eye surgery process.

used, these tissues absorb none of the powerful beam. Instead, all the laser's energy is deposited in the dark absorbing detached retina where it heats the tissue, welding it back together.

a way completely analogous to the isolated atom, the allowed electron transitions involving the absorption of a photon are still from full to empty states. The difference in this case is that the energies of these allowed transitions span a wide continuous range. In the figure these allowed transitions are represented by any of the solid vertical arrows originating in the lower filled red band and terminating in the upper blue empty band, where the higher the energy (or frequency) of the light being absorbed, the longer the vertical arrow. Thus such a solid will absorb light over a wide range of energies, extending from energies represented by the shortest solid arrow shown in the figure to those represented by the longest.

Even more interesting is the concept of forbidden transitions, those shown in the figure as dashed lines. A quick look will assure the reader that there are some energies (i.e., some lengths of vertical arrow) for which no connection from filled red states to unfilled blue states is possible. Specifically, vertical transitions of length less than the vertical width of the forbidden energy gap in the figure will not be able to connect red to blue and hence will not be allowed by quantum mechanics. What this means in the macroscopic world is that light of energy less than the forbidden energy is not absorbed at all in the solid. This is the underlying principle behind the phenomenon of transparency. The reason a diamond is clear is that its forbidden energy gap is so wide that *all* energies of light in the visible spectrum are not absorbed.

Light Emission

Now that we better understand the process of light absorption, we can turn to the concept of light emission, which is at the heart of a working laser. We can understand light absorption as the elevation of an electron from a filled, lower-lying quantum energy level to an unfilled higher one, resulting in the annihilation of an incoming photon whose energy is precisely equal to the difference between the two energy levels. Light emission represents the same process unfolding in reverse. An excited electron in a high level falls in energy to an empty lower-lying level. Although the lower levels are normally all filled, sometimes a number of electrons have been elevated to higher levels, leaving some holes below into which the excited electrons can fall. The falling electron loses energy, and that energy goes into creating a photon equal in energy to the difference between the two levels. The two processes of absorption and emission are in a sense mirror images of one another.

One obvious question arises: How did the electron get in the higher energy level from which it eventually falls in the first place? In our description of the construction of an atom, we started with a series of allowed quantum states and then filled them from the bottom up with the requisite number of electrons needed to make the atom, much the way a bathtub fills from the bottom upward. This arrangement represents the overall atom's state of lowest energy, or what chemists call its "ground state." To return to our bathtub analogy, the bathtub is in its lowest energy state when its water is lying flat in the tub bottom. But if we add

THE SOLID

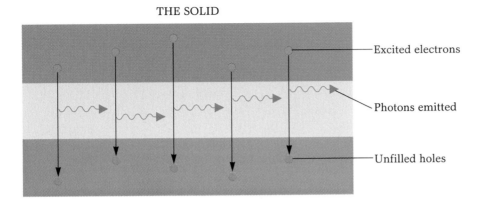

Excited electrons

Photons emitted

Unfilled holes

When an excited electron in a solid jumps downward from a state in the upper band to an empty state in the lower band, it gives up its excess energy by emitting a photon. A similar effect takes place in the isolated atoms of a gas, shown sequentially at the bottom of the figure, except that an electron jumps from higher to lower *discrete* orbital states, again giving up its excess energy as a photon.

THE ATOM

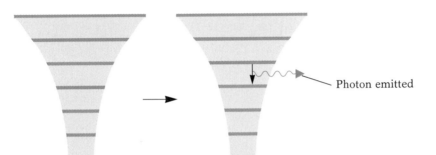

Photon emitted

extra energy in the form of a two-year-old splashing wildly in the water, the bathtub ends up in a higher energy state in which some of the water is flying in the air above the water line! So, too, if we add energy in some way to the ground state of an atom, we can promote electrons to some of the higher-lying previously unfilled levels. Straightforward light absorption can supply the energy, as discussed earlier, or the energy may come from some other external source such as heat or an electric discharge. And just as the water droplets above the tub will eventually fall back into the water at the bottom, so the excited electrons will eventually decay to lower-lying states and emit photons on their way down in energy. This is the process illustrated schematically in the figure above.

The light that arises from such atomic level emission differs from the familiar light emanating from a hot glowing body in a number of crucial ways. The light from a hot body, commonly referred to as "black-body radiation," is distributed continuously across a wide spectrum, as shown in the left graph in the figure on the next page, and is relatively

THE SUN

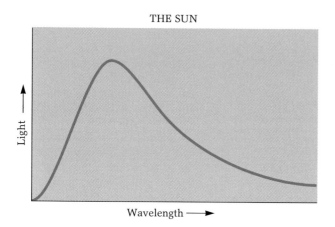

EMISSION SPECTRUM

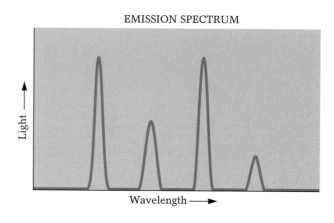

ABSORPTION SPECTRUM

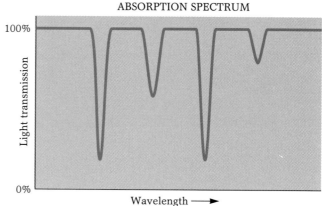

Comparison of the broad spectral radiation from a glowing hot body, such as the sun, with the discrete line spectra of isolated atoms in a gas. The dense collection of interacting atoms in the sun emit light at a continuum of wavelengths characteristic of any body with a surface temperature of 6000 degrees Kelvin, with a peak near a wavelength of 500 nanometers in the green. In contrast, isolated atoms emit light at distinct wavelengths, indicated schematically by the peaks in the top right graph, characteristic of the energies given up as electrons jump from larger to smaller orbits. They absorb light at these same distinct wavelengths as well, indicated by the troughs in the graph on the lower right, as electrons jump up in energy from small orbits to larger ones.

independent of the chemical composition of the body that produces it. Like the light emitted from atoms, this light results from the drop in energy of electrons that had been promoted to higher energy levels, in this case through the application of heat. Because the body is a solid, with its wide bands of allowed energies, and because some of the electrons in a hot body have received enough energy to become completely unbound from the central atoms, giving them the ability to possess any arbitrary amount of energy above this binding threshold, the magnitude of the energy transitions can take on a virtual continuum of values regardless of the makeup of the material that has been heated to the point of glowing. The shape of its spectral output is therefore governed purely by the body's temperature.

The emission of photons from an atomic source, on the other hand, does not rely directly on the temperature of the light-producing object at all, except to the extent that the initial promotion of the electrons to higher-lying states is accomplished by heat. If the promotion is achieved through the use of an electric discharge, as in the case of the familiar storefront neon sign, very little if any heat is required, the truth of the poet's phrase "the cold light of a neon sign."

Atomic light emission differs from the blackbody radiation of a light bulb or the sun in that it is not a continuous spectrum, evidenced in the comparison shown on the facing page. Just as light passing through the interstellar cloud of gas described in the box on page 32 leaves behind its fingerprint of missing energies, so the light emanating from the neon sign consists of a similar fingerprint consisting of a series of emission

Argon

Neon

Nitrogen

Xenon

The emission line spectra of four types of atoms illustrate the unique character of each. The predominance of blue lines in the spectrum of argon, and the predominance of orange and red lines in the spectrum of neon, account for their blue and reddish orange glow, respectively, in commercial sign tubes.

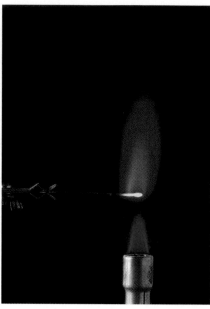

When put into a flame, substances glow with the characteristic colors of their constituent elements, a process that is often used for identification of unknown chemical substances. The three examples illustrate the characteristic yellow color of sodium, the red of lithium, and the blue of rubidium. The colors arise because of the energy differences between electron orbits that are unique to each type of atom.

lines. The only difference is that light is *present* only at the allowed transition energies in the emission case rather than *absent* at those same energies as in the absorption case.

The strongest emission lines give rise to the colors we see in the glowing gas. For example, as seen on page 39, the emission lines for neon are clustered in the orange and red portion of the spectrum, while those of argon are predominately in the blue. Thus, the "neon" sign manufacturer can use those two gases to create reddish orange and blue colors in their glowing "palette." Each chemical element emits its own unique series of light energies when excited, a characteristic used by chemists to perform rapid qualitative assessments of the elements in a given unknown specimen. The "flame test," which you may remember from high school chemistry lab, involves putting a small droplet of an unknown solution in the flame of a burner and noting the color with which it glows. Yellow denotes the presence of sodium, red of lithium, blue of rubidium, and so forth. More sophisticated machines are capable of detecting trace impurities in samples at the parts per million level or better by looking at the intensity and spectral position of such lines.

So we see that light, at least the light that will most interest us in the rest of this book, has its origins in the jumps electrons make in atoms from higher energy levels to empty levels below. To understand exactly how a laser is capable of harnessing this light, we need to discuss one further point. From the moment the electron is first promoted to a higher level, it is in an intrinsically unstable state. That is to say, there exists a more stable state of lower energy, which is precisely the ground state to which the electron later returns. At this point we need to make a distinction in *how* the electron decays back to that stable ground state. Once excited to the upper state, the electron will typically stay in that excited state for a characteristic "lifetime" before spontaneously dropping in energy, and hence "spontaneously" emitting the photon of light. Such a spontaneous drop in energy is the first possible mechanism for emission. This lifetime depends on the details of the atomic orbitals involved, typically ranging all the way from picoseconds (10^{-12} seconds) to milliseconds (10^{-3} seconds).

There is a second mechanism by which this light-producing emission can be triggered, however, first recognized by Albert Einstein in a paper published in 1917. If, while the atom is lingering in its excited state waiting to spontaneously emit a photon, a photon of precisely the same energy as the one the exiled electron will eventually emit passes in its vicinity, it can tickle the atom to prematurely decay and emit its photon. In Einstein's paper, he discussed this process of "stimulated emission" as being the reverse analog of the absorption process. Absorption annihilates a photon while raising the energy of an electron. Stimulated emission creates a photon while lowering the energy of an electron. Furthermore, while spontaneously emitted light can be released randomly in any direction, the emitted photon in this stimulated case is released in precisely the same direction as the photon that stimulated its release, and the peaks and troughs of its waveform exactly match those of the original stimulating photon as well. These particular properties of stimulated emission are crucial for the operation of a laser.

The concept of a laser emerges from an understanding of the consequences of this new kind of light emission. If a situation can be set up in which many atoms are in their excited state, and each photon that is emitted is directed in such a way that it stimulates the emission of additional photons in step with it, one can in principle build up an enormous chain reaction that results in an extremely intense, directed beam of light consisting of photons all of the same energy or color. It is precisely such a situation that is set up within the core of a laser. In the next chapter we will examine more closely how this situation is brought about, as we look at how a laser actually works.

A scientist observes the beam from an argon gas laser. The original single beam has been split into multiple beams by its passage through cubes made of joined pairs of prisms. The laser's multiple beams produce interference patterns inside a crystal that can be used to store data optically.

3

Inventing the Laser

Neon signs are so common that we take them for granted. They have been shining invitingly from nightspots and advertisements since the 1920s, only ten years after Georges Claude first observed the bright glowing colors elicited by passing an electric current through "discharge" tubes filled with neon, argon, krypton, or other gases. These gas discharge tubes were thus available around the time that Einstein published his insights concerning the spontaneous and stimulated processes for light emission. All the pieces were in place to create the laser: the physical apparatus was available, in the form of a neon tube with some significant additions, and a theory existed to point physicists in the right direction. Yet decades passed before scientists stumbled onto the idea, and it was not until 1960 that inventors demonstrated the first neon gas laser. Once understood, however, the elegant physics of the laser received tangible form in a profusion of devices capable of a remarkable variety of uses.

A Trick of Mirrors

We saw in the last chapter that the light that emanates from a simple neon sign has exactly the same source as the light shooting out from a gas laser: electrons that have been excited to higher energy orbitals jump to lower energy orbitals and give up their excess energy by emitting a photon. In the neon sign tube, these excited electrons, unstable in this higher energy state, remain there only briefly before spontaneously dropping to their lower state. The photons they emit can be traveling in any direction and can be observed at any position in the neighborhood of the neon sign tube. These photons are responsible for the sign's red glow.

The same processes that take place in the sign also take place at the heart of a gas laser. In fact, early laser designers often constructed their lasers using a glowing neon tube. Suppose we place a straight neon tube in the center of what will later become our laser cavity. A current made to pass through the neon-filled tube excites many of the atoms within, and the atoms spontaneously give off photons, in arbitrary directions. If our tube were to continue as just a neon sign, the photons would head away from the tube never to return, and we would have a gentle red glow. But to construct the laser, we add another crucial ingredient: mirrors. First we put a mirror at one end of the tube. Most of the photons spontaneously emitted from within the glowing tube head off in directions that do not take them directly to the mirror, and they are lost just like those produced in the traditional neon sign. The actual number of photons produced in a modest 10-watt neon sign is phenomenally high, however: on the order of 10^{19} every second! Eventually one of these randomly directed photons travels directly along the axis of the tube toward the center of our small mirror. Because we have carefully positioned the mirror to be precisely perpendicular to the axis of the tube, the mirror reflects the photon of light back along the line it came from, and it reenters the excited gas.

What happens next is the crucial element that distinguishes our creation from a simple neon sign, making it a true laser. After the photon is reflected back into the tube, it travels through gas filled with many neon atoms that still have electrons in excited higher orbitals waiting to emit photons. In this respect the gas is in a state unlike its normal state, since

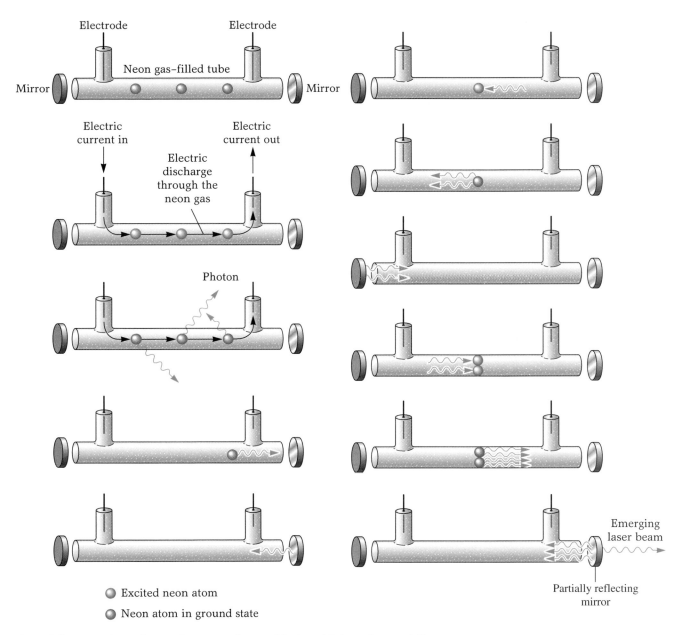

Electrode   Electrode

Neon gas–filled tube

Mirror   Mirror

Electric current in

Electric current out

Electric discharge through the neon gas

Photon

Emerging laser beam

Partially reflecting mirror

○ Excited neon atom

◐ Neon atom in ground state

A straight neon sign tube can become a laser with the addition of a specially aligned mirror at each end. At the left, from top to bottom, atoms excited by an electric current spontaneously emit photons in random directions, eventually producing one traveling directly toward one of the mirrors, from which it reflects back along the axis of the tube. Continuing at the right, top to bottom, the reflected photon encounters an excited neon atom and stimulates a second photon to be emitted. The two travel together as identical twins, reflecting off the second mirror back into the tube. They encounter more excited neon atoms and stimulate additional photons. The right-hand mirror reflects 99 percent of the photons hitting it, while the remaining 1 percent pass through to the outside to become the laser beam.

normally the preponderance of electrons reside in the lower energy ground state, and only a minority in the upper state (those excited by the natural heat energy of any gas existing at a finite temperature). In this case, because the electric discharge current is imparting additional energy to the gas, the electrons in an excited state far outnumber the electrons in some lower state. This inverted state of affairs, required for laser operation, is commonly referred to as a "population inversion."

An excited electron will drop in energy and emit a photon spontaneously, provided the atom is allowed to remain in that higher-level energy state for a decay time representative of that particular atom's orbital structure. In fact, the way that atoms in the neon tube emit light is "spontaneously." But as we saw at the end of the previous chapter, there is a second way this electronic transition can occur, referred to as "stimulated" emission. An excited atom may be approached by a photon of exactly the same color (or, equivalently, the same energy) as the photon this atom is about to emit spontaneously. If the approaching photon comes close enough, it is capable of triggering the electronic orbital transition within the excited atom, thus "stimulating" it to emit its photon before it has time to do so spontaneously. This "stimulated emission" process is at the heart of a laser.

Returning to the simple laser we are "building," the precisely aligned photon that hits the center of our mirror and returns back down the axis of our tube now has plenty of such excited atoms with which to interact, and since it came itself from the same neon atomic transition, it is of just the proper energy to stimulate further such transitions. Here we encounter another crucial feature of stimulated emission: the emitted photon is created exactly in step with the stimulating photon—in other words, the light is coherent. First, the position of the peaks and troughs in the wave it represents, known as the wave's phase, is spatially synchronized with those of the photon that stimulates it, an important feature to which we will return later in the chapter. Second, the stimulated photon travels in exactly the same direction as the photon that stimulated it. It is this collimated nature of the laser beam that allows scientists to measure the distance to the moon to within an inch, since

A photon of just the right wavelength can interact with an excited neon (Ne) atom and stimulate the emission of a second photon, identical to the first in wavelength and aligned with it crest-to-crest as both travel in precisely the same direction.

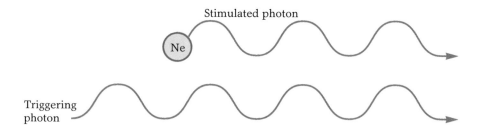

Left: The collimated beam from a laser guides the movements of a bulldozer. The beam is sent from a transmitter mounted atop the tripod (background) in a direction parallel to flat level ground and is rotated around in a 360-degree arc. During its circuit it is picked up by the detector, which is the vertical appendage to the bulldozer's shovel in the foreground. The detector determines the height at which the laser beam intersects the vertical appendage; this measurement gives the precise height of the cutting edge of the bulldozer's shovel. By controlling the cutting-edge height based on this signal, the bulldozer can cut a uniform level plane. Alternatively, if the shovel is moved up and down to conform to the existing landform, the changing heights of intersection, fed into a computer, provide a rapid survey of the contours of the entire field. Right: This darkened view shows the laser beam crossing from the transmitter on the tripod to the receiver on the bulldozer's shovel.

the beam diverges little enough over its half-million-mile round trip to be still detectable on its return to Earth. Surveyors use laser beams in a similar way to make angular measurements from one survey point to the next.

Yet another special property of stimulated emission is its *monochromatic* nature, the fact that it is one single, spectrally pure color. Since the emitted photons all arise from the same atomic-level transition, all are very close in energy. This property of monochromaticity finds a use in the analytic technique known as spectroscopy. Spectroscopic techniques can identify the composition of unknown substances as wide-ranging as atoms, solids, and complex biological assemblages by measuring the spectrum of light coming from the specimen under varying conditions of illumination. The powerful monochromatic excitation of these systems at carefully selected energies made possible through the use of lasers greatly increases the technique's utility.

An industrial laser can cut precise patterns even in sheet metal. Here the 2000-watt beam of a carbon dioxide gas laser, having a wavelength of 10.6 micrometers, is focused to produce very intense heating in a very small area. Precision translators move the sheet metal to allow the focused laser beam to cut specific patterns. The bright streaks streaming from the bottom of the sheet are white hot remnants of metal.

In stimulating the emission of a second photon, the initial photon has in effect cloned itself, and its clone exists at the same region in space, traveling in the same direction, as the original. This fact illustrates clearly that photons are not subject to the Pauli exclusion principle, which dictates that no two electrons can be in the same place with the same energy (or, more precisely, in the same quantum state) at the same time, something we found before to be crucial to the concept of filling up the multiple levels of the atom. In this crucial aspect, photons are fundamentally different from electrons. It is this ability of photons to grow in number and still occupy the same region in space that allows the extremely intense light beam of a laser to form. By directing the already intense beam through the proper optical lens, the entire power of the laser can be focused to an exceedingly small size, on the order of the wavelength of the light itself, giving the very high power densities needed for commercial laser welding or cutting.

So our pair of photons, the original one reflected off the first mirror together with its stimulated twin, continue down the tube to the other end. At the point where the two photons emerge, we place a second small mirror, aligned in such a way that it bounces the photons back into the tube down the same central axis. In this way, the process can repeat itself as each of the two photons tickles another atom to emit a companion photon. Thus a chain reaction begins: the single properly aligned photon becomes two, the two become four, the four eight, and so forth. Eventually, the light traveling in step down the center of the tube far exceeds in intensity the small steady glow of light emitted spontaneously out the sides.

Were the process to stop at this point, we would have indeed created a short segment of intense light within our cavity, but that light, confined to the space between our mirrors, would not be very useful. So the last step is to allow one of our mirrors to let out a small fraction of the light impinging on it, while still returning the bulk of the light to the tube. A mirror formed of thin films of insulating materials can be made partially reflecting: it will, for example, reflect 99 percent of the impinging photons back into the tube, and transmit 1 percent. In this way, a straight stream of light is allowed to emerge out of our laser, nicely collimated into a narrow beam. And if the multiplication process of the chain reaction of photons is strong enough, the gain involved in that process will make up for the light lost in our beam, and the laser will continue to lase indefinitely. In fact, the amount of light emerging will be exactly equal to the amount of light gained in a round trip in the tube.

The process just described has at its heart the amplification of the amount of light through the unique process of stimulated emission. In fact, the name "laser" came originally from the acronym *l*ight *a*mplification by *s*timulated *e*mission of *r*adiation. This process of one photon triggering the emission of a second gives the light in the emerging laser beam the three properties that distinguish it from the light that we see commonly in the world around us: it is collimated, coherent, and monochromatic.

Population Inversion

With a little thought you can see that there is a basic condition that must be met for the laser beam to build up. The system must be excited above the laser threshold condition. There must be more neon atoms in the higher energy state than there are in the lower energy state.

The neon atom in a gas laser has 10 electrons. In its ground state, 2 electrons are in the lowest energy orbital shell, designated $n = 1$, and 8 electrons are in the second-lowest orbital shell, designated $n = 2$. Applying electric current excites the atoms in the laser tube to a higher energy state: the electrons in the electric current bombard the neon atoms and bump their electrons into larger, more energetic orbits, in say the $n = 4$ or $n = 5$ shell. Electrons in one of these states can decay to any smaller, less-energetic state where there is an opening. Electrons in these higher states spontaneously emit the first photons traveling along the tube axis, and because of the high reflectivity of the mirror at a particular wavelength, photons emitted by electrons decaying from the $n = 5$ state become amplified to form the laser beam.

The red laser commonly derived from neon gas has a photon energy of 1.96 eV, which corresponds to a wavelength of 632.8 nanometers. Photons of this energy are generated when a photon of 1.96 eV encounters an excited atom with an electron in the $n = 5$ orbital state. In response, the electron jumps from the $n = 5$ state to the lower $n = 3$ state, stimulating the emission of the new 1.96 eV photon. Then in short order it falls back to the $n = 2$ ground state ready to be recycled. The result of the encounter is the addition of one new photon to the beam inside the tube.

If, however, the photon in the laser encounters an atom with an electron in the lower energy $n = 3$ state, the electron can absorb the photon and jump to the $n = 5$ state. In this case, a photon is lost.

At the threshold for laser action, the number of photons being emitted exactly matches all losses. This means that, for each complete round trip in the optical cavity, the laser would neither gain nor lose any photons. Above the threshold, the laser beam builds up until a steady state is reached, called saturation. The condition for laser action should now be apparent. There must be more electrons in the $n = 5$ state than in the $n = 3$ state; otherwise, more photons will be absorbed than emitted. This condition is called population inversion, since having more electrons in the higher energy state is the opposite of the normal state of affairs for any system.

In practice, additional photons are lost from the window at each end of the tube and at the mirrors, which are not perfect. Therefore, to maintain the laser beam we need a sufficient number of excited neon atoms at all times. The designer of a gas laser must select the density of gas atoms (by choosing the appropriate gas pressure) and the proportion to be excited (by choosing the correct electric current conditions) so that the gas continues in a state of population inversion.

Fundamentally, you can see that it should be a lot easier to obtain inversion conditions if the lower state is not the ground state. It takes a lot of energy to empty a ground state of electrons. If the lower state is not the ground state, however, as is true for the $n = 3$ neon shell, then it is pretty much empty all the time. Hence only a relatively few electrons need to be elevated to the $n = 5$ shell to have the population there exceed that of the $n = 3$ shell. In general, there is one excited state that is easiest to populate to attain inversion. The neon laser is an example of what is commonly called a three-level laser system since three electron orbitals are involved: the ground state, the upper excited state, and the lower excited state.

The gas in a neon laser is almost never pure neon, but contains a healthy admixture of helium. And in fact the neon laser is usually called the helium-neon laser. The addition of the helium gas improves the efficiency of the excitation process. It turns out that electric current can excite helium atoms quite easily to a state that can exist for a relatively long time. Such a state is called metastable because its electrons take their time about jumping back to lower energy orbits. (As explained by quantum mechanics, these electrons are not able to emit a photon in making a jump back to their ground state, and so must give up their energy through other, less efficient means.) However, the energy of this metastable state lies close in value to the energy of the excited $n = 5$ shell in neon atoms. The result is that when, as often happens, an excited helium atom collides with an unexcited neon atom, energy is transferred to the neon atom. The neon atom is left in its excited state, while the helium atom returns to its ground state. Thus the laser is more efficient when helium gas is present to help excite the neon atoms.

The supermarket bar code scanner is perhaps the most familiar application of the narrow beam of a gaseous helium-neon laser. A mirror rocking repeatedly back and forth directs the beam to rapidly scan back and forth in a line. The amount of light reflected back changes as the beam passes over the dark lines and white gaps of a bar code pattern. Another part of the scanner system detects the resultant changing pattern of light reflected back from the code, and a computer interprets the pattern as a particular consumer product.

The Twisting Path to the First Lasers

Adding a pair of mirrors to a neon tube to create a laser seems simple, but the path that led to the invention of the first laser was surprisingly convoluted. The idea of placing mirrors at the end of a glowing neon tube came only after years of struggling with the problem of how to amplify electromagnetic radiation. One of the driving forces behind the struggle was the desire to create a source of light that was both coherent and monochromatic, two key properties of the laser. We've seen some of the fabulous applications unlocked by the ability to generate a strong coherent light beam. Yet light sources of such coherence were unheard of in the 1950s when the story of the invention of the laser began.

Transmitters at that time could already produce radio waves that were quite coherent and monochromatic. The means used then are the same used today: powerful electronic amplifiers drive electrons in the radio transmitter tower up and down an antenna. Since accelerating electrons radiate electromagnetic waves, the electrons oscillating up and down the antenna produce a continuous stream of such waves. The frequency of this up and down motion is carefully controlled by controlling the frequency of the amplifier's output voltage, leading to the output of radio waves at this same very well defined frequency, a monochromatic output. Furthermore, the electrons are driven up and down together, all "in phase" with one another in the parlance of the physicist. As a consequence, each "photon" of electromagnetic radiation—each radio wave—is emitted in phase with thc rcst. In other words, the emitted radiation is highly coherent.

Light sources at the time were much less well controlled. Incandescent sources such as a glowing light bulb filament give off light over a wide range of frequencies, and the random nature of the emission process leads to an almost complete lack of coherence. Yet scientists could not directly extend the "broadcast" method of producing coherent radio waves to the visible portion of the spectrum, for a number of reasons. The height of the antenna is typically on the order of the wavelength of the radiation it broadcasts. That height, which is meters for radio waves, would have to be millions of times smaller to produce the correspondingly shorter wavelengths of light, which are only a fraction of a micrometer (a millionth of a meter). Producing coherent light in this fashion would require a microscopic antenna! And the electronic amplifier to drive the electrons up and down such an antenna would have to operate in the range of 10^{14} cycles per second, three or four orders of magnitude faster than the best electronic amplifiers available even today. Clearly a new approach was needed.

Charles Townes, a professor at Columbia University, invented a device in the mid-1950s that was to provide a way around the dilemma.

Townes designed his device, referred to as the maser (for *m*icrowave *a*mplification by *s*timulated *e*mission of *r*adiation), for use in communications as a microwave amplifier. It consisted of a microwave cavity, of a size just right to fit between its two ends a complete wavelength of the desired microwaves, into which Townes inserted excited ammonia molecules. The excitations of these molecules were not of the same type as those we have been discussing: electrons were not moved up to higher electronic energy levels; rather, the vibrating springs or bonds that hold the nitrogen and hydrogen atoms of the ammonia molecule together shifted to higher levels of vibration. Such vibrational energy levels are much more closely spaced than the electronic energy levels in a gas atom—in fact, energy differences between these levels are more than 10,000 times smaller. Nevertheless, such transitions still possess emission properties similar to electronic transitions: the molecules can drop to a lower vibrational level, emitting a photon either spontaneously or when stimulated to do so by a photon of equivalent energy. But as a consequence of the more closely spaced energy levels, the ammonia molecules emit photons of much lower energy, in the microwave region of the spectrum.

Townes was attempting to create a device that could amplify a weak microwave signal to form a much stronger one. The initial weak signal was passed into the microwave cavity containing the excited ammonia molecules, and this small input of microwave photons stimulated the ammonia molecules to emit additional microwave photons. Within the tuned microwave cavity these photons in turn stimulated others, and the signal grew, producing the desired amplification. Though microwaves can be generated in the fashion of radio waves by driving electrons back and forth using state-of-the-art amplifiers, the maser presented an entirely different approach to the task. In a sense, each excited ammonia atom acts as a miniature transmitter, and the microwave cavity orchestrates these atoms to emit microwave photons together in phase through stimulated emission.

The maser concept provided a radically different approach to the generation of photons. Most important, it defined a way to circumvent the need for an ultrafast electronic amplifier, but could such an approach work at optical frequencies? Atomic transitions between electronic energy levels within atoms emit photons in the visible region of the spectrum, and are thus the perfect candidates for a source of visible photons. The key problem was how to create the resonant cavity. In the case of the maser, Townes physically constructed such a metal cavity to be the size of the microwaves he was attempting to amplify, 1.25 centimeters in his case. Choosing a microwave cavity of that size uniquely selected for that microwave wavelength in the buildup process. But the wavelength of optical light is more than 10,000 times smaller. Scaling the maser down to such a size was out of the question. Another breakthrough was required.

That breakthrough came in a paper published by Townes and Arthur Schawlow in 1958. There Townes and Schawlow theoretically developed the idea of an optical amplifier, based on stimulated emission like the maser, except that the amplifier would be enclosed within a pair of reflecting mirrors that would provide the resonant cavity in which the light would grow. In marked contrast to the maser, however, the mirror cavity would be ten thousand times larger than an optical wavelength. How would it manage to reinforce a wave of a particular size?

Townes and Schawlow correctly surmised that the cavity would, as hoped, select for growth waves of a particular wavelength. These waves would be the ones that, though their electric field value oscillated up and down hundreds of thousands of times during the round trip between the two mirrors, would return precisely in phase. In other words, if a wave started at a given point along the axis at a maximum, it would be exactly at yet another maximum when it returned upon itself after completing a round trip, thus reinforcing itself. Such a stable mode, as it is referred to, would, with the assistance of stimulated emission, constructively build itself up into a strong beam. An added advantage of this axial geometry would be the creation of a highly directional beam, something we have come to associate automatically with the idea of a laser. But in effect this directional property is simply a by-product of the shape of the mode needed to support stimulated growth in the cavity. The key was the pair of mirrors forming the stable mode in the optical cavity. Their insight was later to obtain Townes and Schawlow the Nobel Prize, and as soon as their paper was published, the race was on to create a working device that could create such a beam.

Although Townes and Schawlow had proposed atoms in the vapor phase as candidates for the medium to undergo stimulated emission, other materials were considered by a variety of groups. In particular, Theodore H. Maiman, working at Hughes Aircraft Corporation, had been experimenting with the properties of a solid, ruby. Ruby consists of a host solid, aluminum oxide, into which are added a small concentration of chromium atoms, spread evenly throughout the host. It is in fact these chromium impurities that give the ruby its red color, as pure aluminum oxide is transparent. Maiman surmised that this geometry was entirely equivalent to an arrangement of widely spaced gas atoms in a vapor. The chromium atoms, in this case, are held apart by the aluminum oxide at distances comparable to the distances separating atoms in a gas. More important, his studies had convinced Maiman that the chromium atoms in ruby gave off light at a very high efficiency when excited, making ruby a good candidate for the stimulated emission medium.

Theodore Maiman examines the first laser's original ruby rod in the center of the helical flash tube used to excite it. When in operation, the deep-red laser beam, with a wavelength of 694.3 nanometers, emerged from the top polished facet of the ruby rod.

Maiman began work on the project in August of 1959. For mirrors, he polished the ends of a cylindrical piece of ruby material until they were flat and parallel to one another, then painted them with a highly reflective coating. For stimulated emission to exceed absorption, a majority of the chromium atoms needed to be excited to a higher state, but Maiman could not excite them by applying a simple electric discharge, since that means of excitement works only in a gas. Instead, he surrounded his ruby crystal with an intense flash lamp, much like those used to produce electronic photographic flashes, and its repetitive firing excited the required number of chromium atoms. In May 1960, Maiman achieved a stable red beam emerging from the end of his ruby crystal. He had demonstrated the world's first working example of what was dubbed at the time an "optical maser" and we now know as the laser.

Meanwhile, other would-be laser designers were investigating the use of isolated atoms in the gas phase as sources of light inside the mirrored cavity, and in December of that same year Ali Javan, William R. Bennett, Jr., and Donald R. Herriott at Bell Laboratories announced the creation of the world's first helium-neon gas laser, operating by precisely the scheme discussed earlier. In the years since then a wide variety of

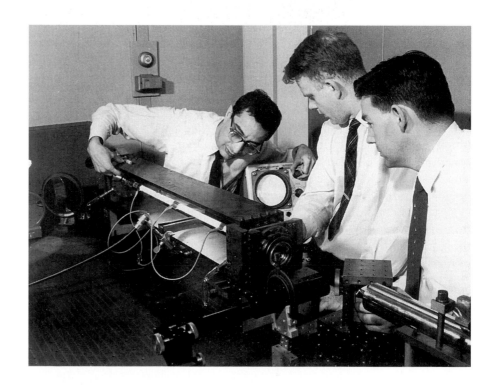

The inventors of the first gas laser — Ali Javan, William R. Bennett, Jr., and Donald R. Herriott — examine the original tube of neon and helium gases. The mirror that is seen at the near end of the tube was the source of the emerging red laser beam, which had a wavelength of 632.8 nanometers.

lasing media have been employed in the construction of lasers, including gases such as argon and krypton and carbon dioxide, solids such as neodymium-doped glass and titanium-doped sapphire, and even liquids. Liquid dye materials are dissolved in a liquid solvent medium, transferred into the resonant cavity with pumps, and excited optically by flash lamps or even other lasers. These liquid dye lasers operate at many wavelengths and can be tuned over a wider range of colors than a solid-state or gas laser.

Looking back, it seems surprising that there was such a long delay between the publication of Einstein's article on stimulated emission in 1917 and the demonstration of the first maser in 1954 and the first lasers in 1960. As we have noted, gas discharge tubes, similar to those used in gas lasers, were in laboratory use even before 1917. Electronic amplifiers, also available early in the century, relied on feedback to modify electronic signals and produce electrical oscillations in a way analogous to the means later used by lasers to amplify light waves. Scientists of the 1920s and 1930s did not try to apply the concept of feedback to light, however, nor did they think about amplifying light using stimulated emission or discharge tubes. Although they were focusing seriously on the fundamental properties of light in both experiments and theory, they did not stumble onto a path leading to lasers until years later. It is interesting to speculate on the reasons for the delay.

Perhaps the primary reason is the ordinary human difficulty in making creative leaps. To conceive of the idea of the laser would have required perhaps too great a leap, especially with all the innovations in basic physics already taking place in the 1920s and 1930s. Physicists of the 1920s, captivated by quantum mechanics, were absorbed in working out the strange and wonderful mechanics of the dual wave/particle nature of both matter and light. The process continued into the 1930s, and even accelerated thereafter, as serious progress was made in the new branches of physics devoted to subatomic particles and the solid state. Perhaps with all these basic innovations in thought and practice physicists already had more than enough to occupy them.

Developments in gas lasers depended on the understanding of gas discharge tubes. Although these devices were not yet completely understood, it seems plausible that someone might have made use of this technology to look for optical feedback effects much sooner than the 1950s, especially since electronic feedback effects were being studied. It is not impossible that the commercial use of neon sign tubes may have discouraged scientists from investigating this area. Scientists are encouraged, even required, to address issues in the forefront of knowledge, and anything commercial may seem stale. The pressure may sometimes even drive researchers away from using the latest advanced technologies. We know of a prize-winning solid-state physicist who for years avoided the use of the latest clean room technologies for attaching electrical connections to semiconductor samples. This was the case even though such technology, highly developed for the fabrication of modern integrated circuit chips, was readily available to him. He preferred to manually apply liquid silver paste, even though the resulting connections were really too big, had poor mechanical stability, and produced poor electrical properties (high resistance). The reason was of course not well thought out, but the point seemed to be to avoid the possible taint of a connection with advanced engineering technology, which was a little too close to commercial practices. It's not unthinkable that a similar attitude toward discharge tubes and neon signs had an effect on basic research in the 1920s or 1930s.

Exploiting Coherence: Gyroscopes and Holograms

No one who has seen a hologram can forget the eerie appearance of these apparitions — suspended in space, they look amazingly real except for the slight hint of transparency and the odd color. Created and projected by lasers, holograms depend on one particular property of laser light, its coherence. The stimulated emission process just described ensures that each photon has the spatial peaks and valleys in its wave at the same position. How wave coherence can be used to produce three-dimensional images is a fascinating story. In fact, coherence is the basis of several of the laser's most interesting applications, including interferometry, used for a variety of tasks ranging from making tiny length measurements to navigating airplanes. Both interferometry and holography make use of the interference patterns created when coherent waves intersect.

Although interferometry was employed prior to the introduction of the laser, it has become much easier to implement and more widespread in its application with the invention of this source of highly collimated and coherent light. As illustrated in the figure on the following page, a beam of laser light is split and sent along two separate paths of different length. When the two beams are recombined, they interfere in a way that is sensitive to the relative shift in their wave patterns, or what is referred to in wave descriptions as their phase. At one extreme, when there is no difference in phase between the two recombining waves, they add together *constructively.* At the other extreme, if the two recombining beams are completely out of phase, then they interfere *destructively,* completely canceling one another. A change in one path length by a mere half a wavelength will cause the interferometer's light detector to

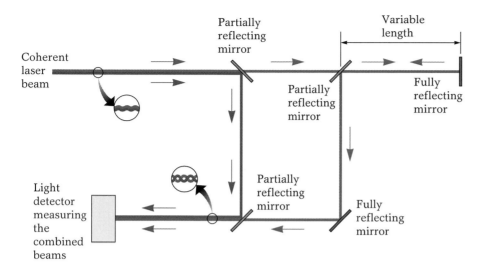

Coherent laser beam

Partially reflecting mirror

Variable length

Partially reflecting mirror

Fully reflecting mirror

Partially reflecting mirror

Light detector measuring the combined beams

Fully reflecting mirror

Schematic of an interferometer. The input laser beam is split by a half-silvered mirror into two equal beams, colored blue and red for illustration. The red beam reflects directly to the lower mirror, which reflects part of it into the light detector. The blue beam follows a longer path to the right, but eventually passes through the final, partially reflecting mirror, where it is coincident with the red beam. Since the two beams have traveled different distances inside the interferometer, their crests and troughs line up differently at the detector. Constructive or destructive interference occurs, depending on the difference in path lengths. By slightly moving the upper right mirror, the path length of the blue beam of light can be varied; moving the mirror only a quarter wavelength changes the blue beam's total path length by half a wavelength which swings the detected signal from maximum brightness to darkness. Its sensitivity makes the interferometer useful for measuring very small length differences on the order of the wavelength of the light used.

swing full scale from bright to no light at all. Since the wavelength of optical light is on the order of a fraction of a millionth of a meter, interferometric techniques are capable of extremely sensitive measurements of length change. Interferometry has a wide variety of applications, ranging from measurements of subtle seismic activity to measurements of slight length changes in the room-size metallic bars that astrophysicists hope will someday detect gravity waves, the echoes of massive black hole collapses in distant galaxies.

One interesting outgrowth of interferometry is the laser gyroscope. The development of inertial navigation, in which such gyroscopes play a leading role, is one of the crucial technological advances in navigation taking place in this century. For centuries, travelers to distant places relied on instruments such as the compass and the sextant, with which a navigator could plot his position by reference to external points such as the Earth's magnetic field or the position of the sun or stars in the sky. But when skies were overcast, they had to turn to dead reckoning. That method called for navigators to calculate where a vessel's speed and direction of travel, recorded frequently in a log, should have taken it, but the accumulated errors inevitable at that time could send a ship hundreds of miles off course. In this century, with the development of the gyroscope, dead reckoning is once again in wide use. It provides the primary navigational guidance for missiles and spacecraft, and is even now the mainstay on commercial aircraft. Modern systems are capable of locating the position of an aircraft to within a few miles after a long transatlantic flight, without taking note of any external reference points at all.

Depending on their phase relative to each other, two waves of the same amplitude and wavelength interfere with each other in various ways. In the top panel, their phases are equal, and they add constructively to yield a wave amplitude that is twice as high as that of either of the original waves. In the middle panel, their phases differ by exactly half a wavelength, and they interfere destructively to produce no light at all. In the bottom panel, their phases differ, but not by precisely half a wavelength, and they sum together to produce a wave of intermediate amplitude.

CONSTRUCTIVE INTERFERENCE

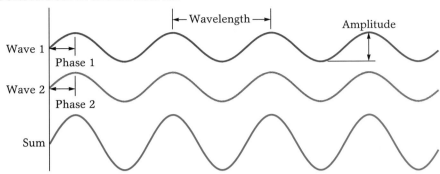

DESTRUCTIVE INTERFERENCE

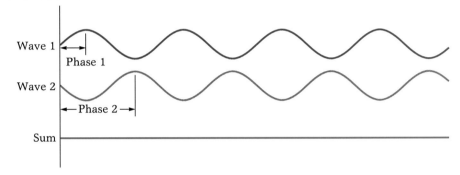

INTERMEDIATE INTERFERENCE

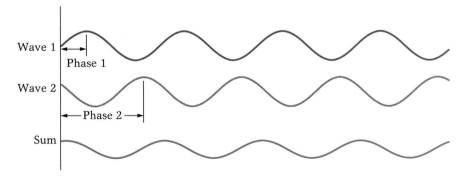

The mechanical gyroscope has a long history. As early as 1914 Elmer A. Sperry developed a gyroscope system called "cruise control" to hold an aircraft in straight and level flight automatically. The gyroscope is a close relative of the child's spinning top, which, as it spins, resists any tendency to change its axis of rotation, an effect due to familiar conservation of angular momentum considerations covered in any high school physics course. It's precisely this tendency that helps make a bicycle much easier to keep from tipping when its wheels are spinning rapidly rather than slowly. The more modern mechanical gyroscope systems provide a platform within the vehicle on which is mounted a set of such gyroscopes pointing along the three spatial directions. These gyroscopes are free to swivel independently of the vehicle—say, an airplane—in which they are mounted. As the aircraft changes course, the platform swivels within its mount in such a way that it holds its direction constant even as the plane turns. A precise measurement of the change in angle provides an attached computer with an exact change in course heading. Furthermore, within the platform are mounted accelerometers that detect any changes in speed along each of the three directions. This information on speed and heading is combined in the associated computer to form a complete record of the entire flight, thus determining the present position at any given time.

Engineers are concerned about the reliability of mechanical gyroscopes because the moving parts necessary for the rapidly spinning rotors may gradually wear out. A laser-based technology, founded in particular on the coherence properties of laser light, has now begun to supplant these mechanical devices. Imagine making our gas laser not a single long tube, but a combination of four glowing sections forming a square, with mirrors at each of the corners. In such a configuration, known as a "ring laser," we have two separate beams of light traveling

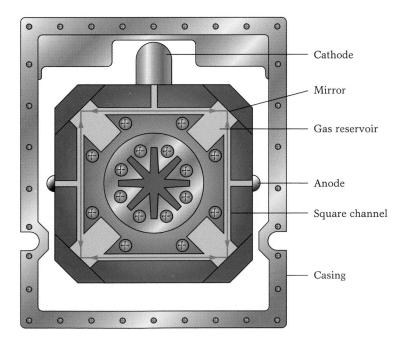

Cathode

Mirror

Gas reservoir

Anode

Square channel

Casing

This ring laser is made from a single block of glass in which channels have been hollowed out and filled with a gas mixture. Electric current flows between the cathode (negative electrode) and the anodes (positive electrodes) to excite the gas and produce light. The light reflects at right angles off a corner mirror from one channel to the next, setting up the conditions for efficient stimulated emission of light along the four connected channel axes, and forming the optical cavity for a ring laser. This is the basic structure of an optical gyroscope. In operation, optical interference between light beams traveling clockwise and counterclockwise around the square provides the readout.

around the square in opposite directions, instead of a single beam of light reflecting back and forth between two end mirrors.

If we tap these two beams out at some point along the square and combine them at an optical detector, the peaks and troughs of each will add in either a constructive or a destructive way, depending on their relative phase. As long as the ring laser remains stationary and nothing changes, this combined signal will remain constant. Now imagine that we start to rotate the square. At this point the well-known Doppler effect comes into play: the light wavecrests arriving at our detector from the beam moving toward it as the ring rotates will begin arriving faster, just as those from the other beam will arrive more slowly. This difference in the frequency at which peaks from the two beams arrive at the detector will cause the combined beam to go through a series of changes from constructive to destructive interference and back. Our detector will begin to see a sequence of changes from light to dark to light, and so forth. The frequency of this "winking" depends on the speed at which we rotate the ring. If we turn it twice as fast, the winking occurs at twice the rate.

All we need now to complete the laser gyroscope is a computer that measures the rate of winking, and we have a way of determining changes in direction without resorting to any moving parts whatsoever. This is the basic principle behind commercial optical gyroscopes, which have now become standard issue on modern aircraft such as the Boeing 757 and 767.

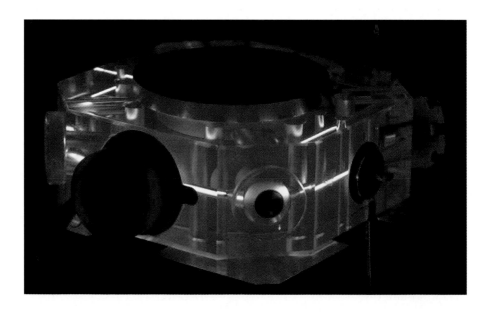

View of the ring laser shown on the previous page, set up to function as an optical gyroscope. The near corner mirror can be clearly seen reflecting the red laser beam to help form the ring-shaped optical cavity.

The coherence of laser light perhaps finds its most dramatic application in the hologram. Now familiar from uses as varied as the dove logo on Visa credit cards to the heads-up three-dimensional displays in modern aircraft cockpits, a hologram is a way of capturing and reprojecting a three-dimensional image of an object. Its recording of the light from the object is dramatically different from that of a photographic image. Conventional photography is best understood by examining the process used in a pinhole camera. As illustrated on the next page, light from each point on an object being photographed travels in a straight line through the pinhole to a single corresponding point on the photographic plate. The plate records the intensity at each point, and from this recording a complete copy of the object is captured.

Substituting a lens in place of the pinhole, as is customary in photography, doesn't change the image at all. The lens is simply an optical element constructed in such a way that it bends all the rays emanating from a given point on the object, no matter which part of the lens they hit, in such a way that those rays all focus on the same given point in the image. This the lens does for each point on the object, and although the image does not change at all in its shape, it becomes more strongly illuminated since each point is now the sum of many rays.

What is lost in the photographic image, however, is any three-dimensional information. Its disappearance can be traced in part to the fact that conventional film does not record all features of the light hitting it. Light is characterized both by its intensity — how high the peaks and troughs are relative to one another — and by its phase, or the relative position in time of the peaks and troughs. We've already seen how this phase information can lead to a number of interesting interference effects. To completely capture the entire light field of the object emitting or scattering the light, one needs to capture both intensity and phase information. A hologram does just that.

Photographic film, consisting of a thin layer of a light-sensitive chemical, can record intensity variations, as varying amounts of converted chemical in the film emulsion layer, but it records nothing about phase. Hence the limitation of a conventional photographic image. We would like to place a photographic plate between the viewer and the object and record at every point on the emulsion both the intensity and phase of the light traveling toward the viewer. If we could successfully do this, we could later illuminate the plate in such a way that the light will propagate onward from the plate toward our eyes just as it would if the object were present, even though it may have been removed long before. The challenge is to suspend in time the full light front with all its intensity and phase information.

The fundamental idea behind a hologram is to illuminate the object and the plate directly with a plane of coherent light, as illustrated in the

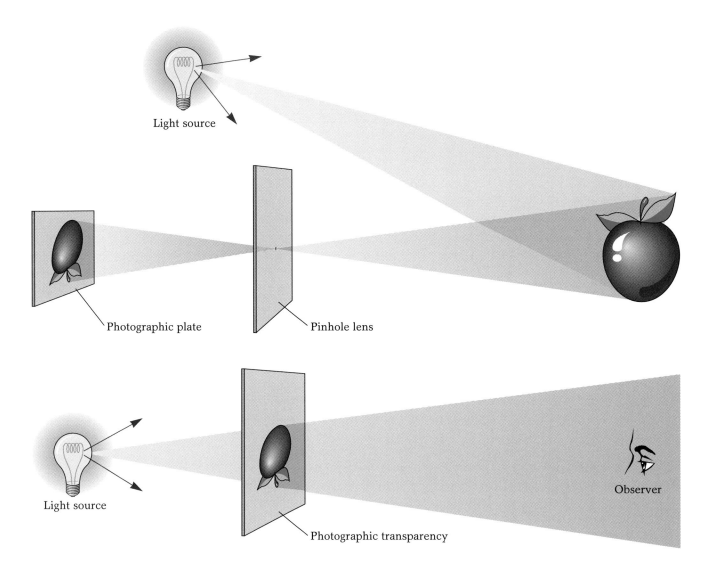

The operation of a pinhole camera. Light scatters off an object, the apple at the right, and some of the light rays pass through a tiny pinhole in an upright opaque screen. The pinhole insures that each point on the apple illuminates only a single point on the photographic plate beyond. Upon striking the plate behind the screen, the rays form a dim but clear image, which can be recorded photographically as a pattern of varying light intensities. The image is sharp only for very small pinhole sizes, limiting the total intensity of light arriving at the plate. Modern cameras increase the light level at the plate, and film, by enlarging the pinhole opening and placing a lens in it to maintain a focused image, still just a pattern of varying light intensities. In the lower part of the figure a light source illuminates the photographic transparency formed in the first process, allowing intensity variations recorded on the slide to be transmitted to an observer's eye.

figure shown below. Unlike the photographic case, now each point on the plate receives light reflected from *every* point on the object, just as an imaginary plane placed there in space would. Since the lighting is coherent, the rays all combine at each point in a unique way indicative of the phase of the light wave at that imaginary plane. The trick to capturing

RECORDING STAGE

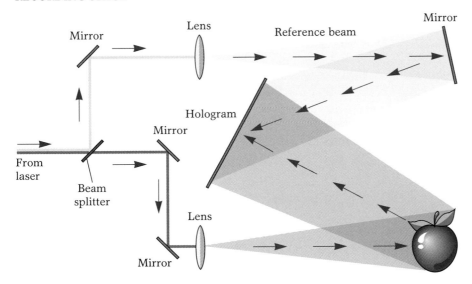

RECONSTRUCTION STAGE

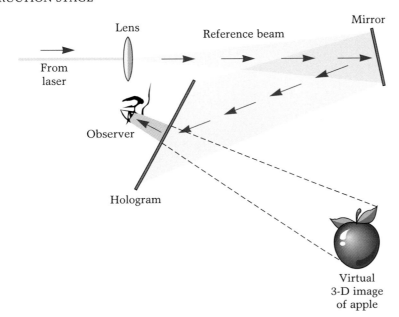

Laser beams can be used to create a hologram. A laser beam entering from the left is split into two parts by a partially reflecting mirror acting as a beam splitter. The lower beam then illuminates an object and scatters toward the plate on which the hologram is to be formed. Simultaneously, a reference beam of coherent laser light, the upper beam, is sent directly to the plate, where it interferes with the light from the object. The interference pattern produced on the photographic plate stores both intensity and phase information about the light arriving from the object, giving a more complete record of the object's appearance. An observer can view the object recorded in the hologram by illuminating the film with a laser beam. The light scattered off the hologram toward the observer's eyes has the same intensity and phase information as the light that would have come through the window of the film from the real object, producing a realistic three-dimensional view of a virtual object which no longer actually exists in space.

this combined phase information on the film is the use of a second direct coherent beam, extracted from the same laser source, called the reference beam. This second beam is projected directly onto the plate, where it interferes with the waves arriving from the object in a way that can be captured photographically: dark spots appear where the two beams interfere destructively and bright spots where they interfere constructively. It in effect converts the phase information at the plate, unrecordable by the film, into intensity information that is recordable.

Later, when the object has been completely removed, anyone can reconstruct the entire plane of light coming from it, complete with the necessary phase information, by simply illuminating the plate with a coherent laser beam, as shown at the bottom of the figure on the preceding page. To the observer a virtual object appears, looking identical to the original. The observer's eyes each see a slightly different view, since the impinging beams are identical to those that would have reached his or her eyes through the imaginary plane, and the object can even be viewed from different angles as long as they are small enough to pass

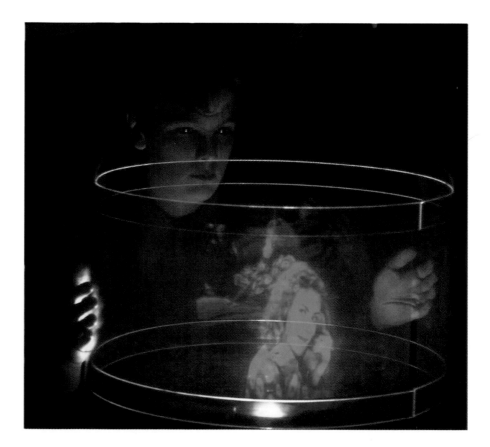

Holograms are favorite exhibits at museums; the one seen here is from the Museum of La Villette in Paris. Arranging the hologram material within a cylinder provides an especially striking three-dimensional view.

through the part of the imaginary plane covered by the original photographic plate. This is the origin of the three-dimensional effect we commonly associate with holograms.

This novel concept dates back to an original proposal made by Dennis Gabor in 1947, well before the invention of the laser, for use in the field of electron microscopy. It was Gabor, in fact, who coined the term "hologram," combining the Greek *holos* and *gramma,* meaning "the whole message." Even at that early date, physicists had achieved coherence in the beam of electrons used to image samples in an electron microscope, despite their inability at the time to create a source of strong coherent light. On the other hand, lenses for bending the electrons, the equivalent of a camera lens, were very primitive compared to optical lenses.

Gabor's idea was to form a holographic image by having the electron beam emanating from his sample interfere with a reference beam tapped from the starting electron beam, all without the use of lenses. Then, once the photographic plate was developed, it could be illuminated with light and further magnified using high-quality optical lenses, readily available at the time. Although the technique was an interesting way around the use of lenses, the later development of extremely good electron lenses negated its utility in the microscopy field, but shortly after the invention of the laser in the early 1960s, a team of scientists, Emmett Leith and Juris Upatnieks, at the University of Michigan, employed the technique using that newly discovered coherent light source and thus created the field of optical holography.

All the lasers discussed in this chapter rely on isolated atoms or molecules to serve as independent sources of photons for the lasing action. In the case of the gas laser, the atoms are kept independent from one another by the large distances that separate atoms in the gas phase. In the more dense solid and liquid phases of other lasing media discussed thus far, these distances are achieved by diluting the optically active atoms or molecules in an optically inert host, as the chromium of a ruby laser is diluted in aluminum oxide or the dye molecules of a liquid dye laser are diluted in their host solvent. There is, however, another class of laser, the semiconductor laser, that does not depend on independent isolated atoms for its source of light. Rather, the properties that lead to the creation of light are the result of a complex interplay between all the atoms in the semiconductor solid. The novel characteristics of these lasers allow them to be made exceptionally small and efficient; they are the lasers that will be at the heart of the coming Age of Optics. To understand this technologically important class of lasers, we turn in the next chapter to the basic properties of solids and the atoms of which they are made.

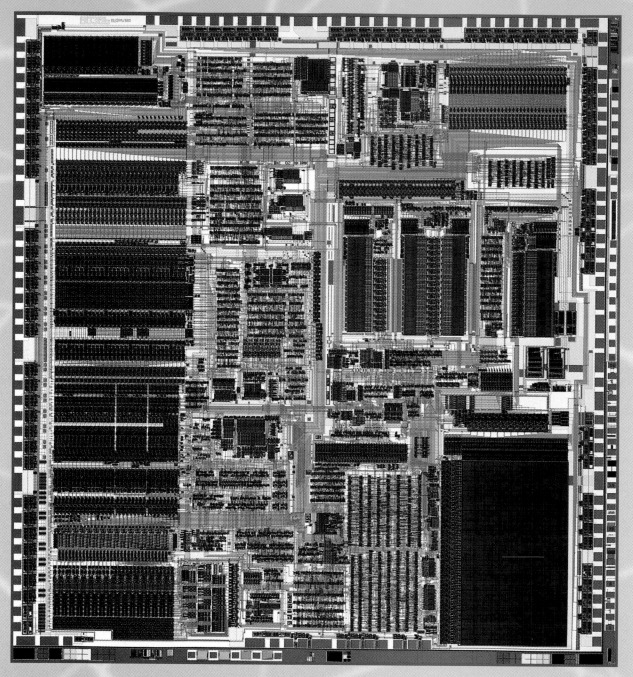

A diagram of the Intel 386 (TM) microprocessor chip, which served as the basis of an entire generation of IBM PCs in the late 1980s. Such a computer-generated diagram provides an overall map of the various layers of micropatterning required to form the layers of doped silicon, insulating oxide, and conducting metal that make up the final integrated circuit. The semiconductor technology that created this chip would also spawn a new class of lasers.

4

Bonds, Bands, and Semiconductors

In gas lasers as in ruby crystal lasers, the photons that make up the laser beam have their source in electronic transitions taking place within individual atoms. Each atom acts as an isolated system — each is coupled to other light-emitting atoms only by the lasing process itself, as the photons released by one atom tickle other atoms to emit in phase. The key properties of the atoms themselves, such as the energy levels and electronic orbitals, are still basically properties of each isolated atom, unaffected to any significant degree by the surrounding material, whether gas, liquid, or solid.

But not all lasers are made from materials that act as assemblages of isolated atoms. One of the most technologically important materials in widespread use today is the semiconductor, which now serves as the basic building block of modern computers, digital electronics, and the other marvels that have led some to refer to this era as the "Semiconductor Age." Hence it may not come as a surprise that an important class of laser, the semiconductor laser, makes use of these exciting new materials as well.

Given the phenomenal miniaturization achieved by semiconductor technology, it is no wonder that the research community has turned to semiconductors to form miniaturized rugged lasers. Lasers smaller than a grain of salt are now in routine use in applications such as the readout device of a compact disc player, while scientists at the frontiers of research have created devices a millionth of a meter on a side. Lasers at these incredibly small volumes can turn their light output on and off at phenomenally fast rates, so that some modern devices are capable of emitting tens of billions of pulses a second. This ability has opened the way to the rapidly expanding implementation of high-speed fiber optic communications.

Scientists have been studying materials that we would call semiconductors since the 1830s. In 1833 the British physicist Michael Faraday noted that silver sulfide became a better conductor of electricity as the temperature rose, whereas most conducting substances (and all true metals) become worse conductors when heated. We recognize today that this "normal" behavior of metals arises from the fact that as we heat the material, its atoms begin to vibrate more vigorously, causing them to get in the way of the smoothly moving electrons carrying the electric current through the metal. Yet Faraday found that one class of conducting materials, of which silver sulfide was one, behaved in the opposite way on heating. Over the next hundred years physicists continued to notice that some materials, thought to be metals, had additional peculiar electrical properties. The element selenium, for example, was observed to undergo an appreciable change in conductivity when illuminated. These peculiarities were noted with interest, but physicists had no idea what to make of them.

Scientists of the nineteenth and early twentieth centuries did not realize that they might be dealing with a fundamentally different class of material. They classified solids as either metals, which conduct electricity, or insulators, which do not. Although at best mediocre conductors, the materials with the curious electrical properties noted above were categorized as metals, and not unreasonably so. Although all true metals conduct electricity, some are much better conductors than others. Metals vary widely enough in their properties that as a class they seemed well able to encompass a few offbeat compounds and elements of slightly lower electrical conductivity. Physicists did not begin to understand the unique nature of the materials we now call semiconductors until the arrival of quantum theory in the 1920s helped them to see how atoms and electrons behave in a solid. They found that the atoms in a semiconductor solid interact in exciting and complex ways.

The Bond in Chemistry

Atoms in a gas and atoms in a solid exist in profoundly different environments. Recall from Chapter 2 that, for the atoms in a gas laser, the size of the electron orbitals in a single atom, which in turn defines the effective "size" of the atom, is a few angstroms (10^{-10} meters). In the optical cavity of a gas laser, atoms measuring a few angstroms are separated by many tens to hundreds of angstroms. So to all intents and purposes, each atom is effectively isolated and can be treated as a separate system. This view leads to the classic model of a gas as an assemblage of Ping-Pong-ball-type objects whizzing past one another, only occasionally colliding with other atoms, and then only for brief periods. An atom in a gas spends most of its existence in isolation.

In such an environment, each atom's orbitals and corresponding energy levels are distinct to that atom and are not influenced by equivalent properties of other atoms in the gas enclosure. Hence a mixture of helium and neon in a helium-neon laser cavity consists of a combination of *some* atoms with the energy levels of helium and *some* atoms with the energy levels of neon. *None* of the atoms have energy levels lying somewhere *between* those of helium and neon. Similarly, chromium atoms, which produce the light of a ruby crystal laser, are distributed evenly throughout the ruby crystal and are spaced far enough apart that they, too, can be treated as separate atoms isolated from one another.

But the picture becomes significantly more complex when the atoms that emit the light are packed tightly together and become the major constituent of a solid. The atoms in a solid are densely packed—three orders of magnitude more densely than the atoms in a gas. The spacing between these atoms now becomes on the same order as those few angstroms that are taken up by an individual atom's electron cloud. This leads to profound effects that are the underpinning of much of the discipline of chemistry. As we briefly discussed in Chapter 2, the electron originally bound to one atom begins to be free to wander into parts of space where it is closer to other nuclei, and the whole concept of an electron being a part of a single atom's electron cloud begins to crumble. We can no longer treat the atoms as isolated entities; instead, we need to look at their interactions with one another. We must treat properties such as allowed electron energies as properties not of the isolated atoms, but of the solid as a whole.

A schematic representation of the three 2p electronic orbitals on the same atom. Each orbital consists of a pair of opposing lobes lying along one of the three principal axis directions shown in the figure. Since each of the three orbitals can individually accommodate one "spin up" electron and one "spin down" electron, the entire set of orbitals can hold a total of six electrons when completely filled.

This transition from a group of isolated atoms to a unified, interrelated solid can be thought of as occurring continuously as we bring the atoms closer and closer together. There are a couple of ways for us to visualize what is happening as this spacing shrinks and the interaction between atoms becomes significant. One can view the result from the perspective of either a chemist or, as we shall see later in the chapter, a solid-state physicist.

The chemist focuses on the attractive forces that form a "bond" between two atoms as they closely approach one another. In the chemist's world, electrons provide the "glue," the interaction that binds the entire solid together. The key to this binding together is the tendency of electrons to rearrange themselves into lower energy configurations. Two closely packed atoms can arrive at a lower energy state by reconfiguring their electron clouds—for example, by forming orbitals that bridge the two atoms or alternatively accomplishing the net transfer of electrons from one atom to the other. The result is that the two atoms stay bound to one another, and in the chemist's parlance, a *chemical bond* is formed between the two atoms.

A chemist looks at the orbitals of a typical atom, such as the three *p* orbitals depicted schematically in the figure on this page, as a map of sorts showing the directions along which the atom extends. In the case of the three *p* orbitals found in a single shell, there are three such directions, extending along the x-, y-, and z-directions. If these three orbitals are completely filled with the two electrons per orbital allowed them, then the electronic shell is said to be filled and the atom is unlikely to bond with other atoms.

The opposite is the case, however, if each of the three *p* orbitals contains only a single electron. Such orbitals are in a state ready to react with other atoms with similarly half-filled orbitals. The chemist views the directions in which the electron cloud lobes extend as the directions along which the atom will first interact with adjacent atoms as the atoms become more closely spaced. As the lobe of one atom's electron cloud begins to overlap that of another atom, the electrons associated with each overlapping orbital begin to feel the attraction of the other atom's positively charged nucleus as well as their own. As a result, they start to form a combined orbital concentrated in the space between the two atoms. The electrons in this orbital are now associated not with one or the other of the two atoms, but rather with the new chemical "bond" that has now formed between the two. This combined orbital is anchored to both the starting nuclei, and it in turn ties the two nuclei together.

Not all bonds are identical; the distribution of the electrons in the space between two bonded atoms varies with the elements involved. The nature of the bond formed by an atom will profoundly influence

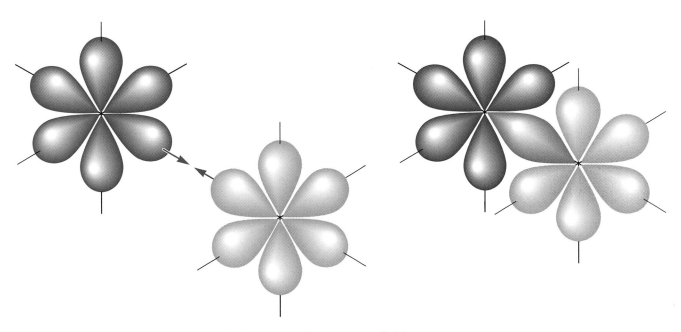

A schematic representation of the *p* orbitals on two adjacent atoms (left) combining to form a chemical bond (right). The electron cloud that becomes the new bonding orbital, represented in orange, becomes concentrated in the space between the two bonding atoms. This new orbital contains the electrons, from both starting atoms, that were associated with the two overlapping orbitals. For the sake of clarity, we have represented the electrons bound separately to each of the starting atoms as red and yellow respectively, and those shared by both atoms once the atoms overlap as orange, a mixture of red and yellow. The remaining red and yellow portions of the bonding orbitals, the lobes facing away from the two atoms along the line of the bond, actually shrink and become reshaped in the bonding process, but for simplicity we neglect this effect here.

whether atoms of that type, included in a material, will help give that material the unusual properties of a semiconductor. An atom's bonds are crucial to the light-emitting ability of a semiconductor laser.

It was noted more than a century ago that certain groups of elements share common chemical properties. For example, the group referred to as the noble or inert elements (He, Ne, Ar, Kr, Xe, and Rn) all exist as gases at room temperature and rarely react chemically with other elements. Similarly, salts are often formed by mixing equal parts of an element from the alkali metal group (Li, Na, K, Rb, and Cs) with an element from the group of halogens (F, Cl, Br, and I). Each element has an atomic number associated with it indicating the number of positively charged particles (protons) in its nucleus. This number, shown immediately above the element's chemical symbol in the familiar periodic table on the following page, is also the number of electrons the atom has

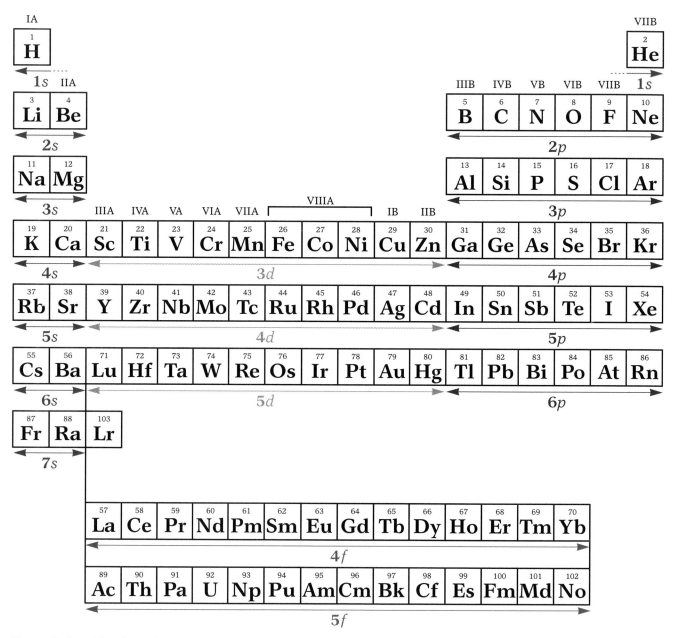

The periodic table of the chemical elements, showing each element's atomic number above the symbol for the element. Arrows underneath the rows indicate the outermost electronic shell being filled. Some of the *d* and *f* orbitals can, in the case of some atoms, fill after the next higher shell number s-levels in that atom.

when it is in its neutral uncharged ground state. The arrangement of elements in the periodic table, and particularly what it tells us about these electrons, provides vital insight into bond formation.

When the sequence of known elements is placed in order of increasing atomic number, an interesting pattern appears, first noticed in the nineteenth century. Elements from the same chemical groups (such as the inert gases, alkali metals, and halogens) appear in this list at regular intervals. The periodic table is arranged to reflect this pattern: the atoms are placed in sequence by increasing atomic number, but the table breaks to a new row at strategically chosen places to form vertical groups of atoms with similar properties (such as the inert gases, the halogens, and the alkali metals). Thus, all the elements with similar properties appear in the same column, or group, designated by a Roman numeral placed at the top.

The electronic orbital theory discussed briefly in Chapter 2 is the underlying mechanism that explains this periodicity, and it is this quantum theory, proposed in the first quarter of the twentieth century, that determines the underlying bonding configuration. Without going into the details of how the shapes and energies of the individual orbitals are derived in quantum theory, let's examine the energy levels of each of the possible electronic orbitals such as those shown on page 25. The levels are displayed in the figure on this page, organized horizontally by their shell number (referred to as their "principal quantum number") and vertically by the energy needed to elevate an electron into that given level. The letters s, p, d, and f indicate the physical shape of the orbital, cataloged earlier on page 25. The number of electrons (e⁻) each level holds when full is indicated to the right of each level. There can be multiple orbitals of a given type, each with a different spatial arrangement—like the three p orbitals shown on page 70—but all with the same energy. Since each orbital can accommodate 2 electrons, this means a complete p orbital system for a given principal quantum number can contain 6 electrons in total. This accommodation level rises from 2 to 6 to 10 to 14 as we move up the series from s to p to d to f, since the more complex higher-lying orbital shapes have a correspondingly larger number of energetically equivalent orbitals (1, 3, 5, and 7 for the s, p, d, and f orbitals respectively).

The array of levels shown in the figure on this page tells us where nature *allows* us to stash our electrons. The number of electrons we *need* to stash, for a neutral uncharged isolated atom, is equal to the number of positive charges in the nucleus, or the atomic number of the atom. To obtain the lowest energy configuration for these electrons, what is referred to as the atom's ground state configuration, we must fill the electrons into the states shown in the figure, lowest energies first. It's like filling a bathtub with water: the energy levels fill from the bottom up.

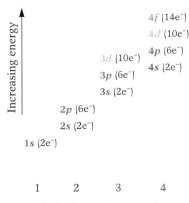

The various atomic orbitals each have a specific energy, represented by their height in the diagram, that results from the average spacing between the positively charged nucleus and the orbital's negative electron cloud. Numbers to the right of each level in black indicate the total number of electrons that that set of orbitals can accommodate. Orbitals are grouped into shells, indicated by their principal quantum number at the bottom of the figure; energy levels in each successive shell are higher on average than those of their predecessor. As one moves through the periodic table, the orbitals fill in order of increasing energy: the 1 levels fill before the 2 levels, and the 2 before the 3, but, because of its lower energy, the 4s orbital (K and Ca) for example fills before the higher-lying 3d (Sc, Ti, V. . .).

This entire process is a reasonably complex one, but if we step back from the details, we can see an important pattern emerging. If we think of the process of filling these levels, the "water level" doesn't rise continually as in our analogous bathtub. The "energy level" in the atom stays relatively fixed as we fill an entire shell with electrons and rises when we are forced to go to the next shell. Our "bathtub" fills in steps, associated with each allowed energy level. If we take the sum of all the energies of the filled levels and divide this quantity by the total number of electrons, we arrive at the average energy per electron. This average represents a sort of figure of merit telling us how relatively stable the atom is. This stability figure of merit takes a jump whenever we begin filling the next higher shell, which means the completely filled shell, just before this jump, is relatively quite stable. Thus, the atom is most stable, in terms of the average energy expended to fill the levels, when it has just completed filling an entire shell. This result is the key we need for the rest of our discussion of bonding: the preferred lowest-energy electronic orbital configurations (or bathtub fillings if you like) are the completely filled shell configurations.

With this in mind, let's look again at the periodic table on page 72. Beneath each row of elements are shown the orbitals of the figure on page 73 that are being filled as we move from one element to the next. Note that the periodicity is directly related to the complete filling of given levels. In particular, the elements having completely filled p levels—the arrangement resulting in the lowest energy configurations—correspond precisely to the Group VIIIB inert gases. They *all* possess completely filled p shells, and hence are low in energy and quite stable. In fact, the various numbers of electrons that correspond to the filled shell configurations of these inert gases (2, 10, 18, 36, 54, 86) are "magic" numbers: through chemical bonding with other forms many of the other elements in the periodic table achieve outer shells that also contain a "magic" number of electrons.

To see how this is done, let's return to our two atoms as their electron clouds begin to overlap. We can now examine how the electron-sharing arrangement in the overlapping "orange" bonding orbital between the two atoms is determined. Suppose that the two starting atoms are dissimilar, with widely differing attractive power for the bonding electrons (a property that we will refer to here as the atom's "attractive power" for electrons). In this case, the two electrons shared by the two atoms will gravitate to the atom with the higher attractive power. The bonding orbital will become skewed toward that atom, which means that the two electrons in that orbital will spend more time near that atom. Such an arrangement effectively robs one atom of an electron, causing it to have a net positive charge (such an atom is referred to by chemists as a positively charged "ion"). At the same time, a net negative charge forms on the atom to which the extra electron has migrated (that atom becomes a negative "ion.")

The solid form of sodium chloride, the common table salt we use in preparing food, is just an assemblage of such positively charged Na^+ and negatively charged Cl^- ions. An atom of sodium, which has 11 electrons, is "happy" to lose an electron because its remaining electrons fall into the stable 10-electron structure of a filled shell, that of neon. On the other hand, an atom of chlorine, with 17 electrons, needs just one more electron to achieve the stable 18-electron configuration of argon. In general, atoms on the right side of the periodic table can reach the stable configurations by grabbing extra electrons and hence have a high attractive power. Similarly, elements on the left side must shed them to reach the stable number and hence hold onto their electrons only weakly; these elements have a low attractive power. These differences in where sodium and chlorine lie with respect to the nearest stable Group VIIIB configuration are in fact the reason for the large difference in attractive power that makes the shared charge of the bond gravitate away from the sodium and toward the chlorine. Once ionized in this fashion, the positively charged sodium atom is then strongly attracted to the negatively charged chlorine atom, and the result is the formation of an ionic bond between the two atoms.

Atoms can lose more than a single electron, forming ions with higher positive charges of $+2$, $+3$, and so forth. But since the additional electron is being removed from an already positively charged ion, it will be more strongly attracted to the atom it is leaving. As a consequence, the work to remove these additional electrons increases significantly with increasing ionization number. The same is true of adding electrons to an ion that already has extra electrons—more energy has to be expended. Atoms near the middle of the periodic table would have to ionize many times in order to arrive at the "magic" stable electron numbers and thus bond ionically like sodium chloride. Often the energy required to achieve this much ionization is more than the energy gained by reaching one of the stable filled-shell configurations. How does bonding proceed in such a case?

The bonding process is quite different for atoms that lie in the middle of the periodic table, such as silicon with 14 electrons, midway between the magic numbers of 10 (neon) and 18 (argon). These atoms bond instead by *sharing* electrons with their neighbors. This is the type of bonding that takes place in the case of silicon and the other semiconductor materials.

This sharing process involves the formation of a bond in which the two electrons, one contributed by each atom, gravitate not toward one or

A crystal model of common table salt, sodium chloride (NaCl), an example of an ionically bound solid. The solid is held together by the attraction of the positively charged sodium atoms (Na^+) to the negatively charged chlorine atoms (Cl^-). This spatial arrangement of atoms maximizes the number of attractive ionic interactions between the positive and negative ions.

1

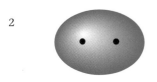

2

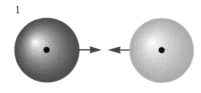

A schematic depiction of how a chemical bond forms between two hydrogen atoms. The two small blue dots represent the nuclei of the two hydrogen atoms being brought together. The shared electronic orbital represented in green below forms the covalent bond between the two original "blue" and "yellow" atoms.

the other of the atoms as in the case of Na—Cl, but instead to the space directly between the two atoms in a more equal sharing as in the process already depicted schematically on page 71. There both atoms "share" the two electrons. Such a sharing arrangement forms what is called a "covalent" bond. Each atom gets to "count" *both* electrons in its quest to achieve a stable filled shell.

The simplest case of a covalent bond is the bonding between two hydrogen atoms, each with a single proton in the nucleus and a single electron in a $1s$ orbital. Referring to the periodic table again, we find the closest stable filled shell (Group VIIIB) is helium's, with two electrons. When the two hydrogen atoms bond, therefore, the two shared electrons (in fact the only electrons in the case of hydrogen) both enter an orbital centered on the midpoint of the line joining the two atoms. In the process the spherically symmetric $1s$ orbitals shown in the upper panel of the figure on this page become deliberately distorted along the direction of the bond to enhance the electronic overlap. A strong H—H bond results. *Both* hydrogens "think" they now have two electrons nearby, and hence both reach the stable lower-energy, filled-shell configuration of two. The stability of the arrangement is the reason hydrogen gas almost always exists as the bimolecular molecule H_2. And it is the reason that the breaking of this extremely strong bond in the burning of H_2 gas releases a large amount of energy. Consider the intense heat of a quartz blower's hydrogen-oxygen torch or the disastrous consequences of the sudden release of this energy during the explosion of the *Hindenberg.*

Let's look at an example more relevant to semiconductor lasers: a covalently bonded semiconductor solid. The simplest semiconductor, although not one useful in a laser, as we will see in a later chapter, is silicon. Silicon is in the fourth column of the periodic table (Group IVB), which means that it has 4 more electrons than the next-lower stable filled shell of 10 represented by neon, and 4 fewer electrons than the next-higher stable filled shell of 18, represented by argon. To achieve a filled-shell configuration of 10 or 18, a silicon atom would have to lose or gain 4 electrons. Ionization of four electrons is far too costly energetically, so some form of covalent bond is necessary.

From the chemist's point of view, a silicon atom has four electrons to contribute to the bonding process. When about to bond, it reconfigures the orbitals of those four electrons into an arrangement with four different lobes, equally spaced one from the other, as shown in the figure at the top of the following page. This arrangement allows the atom to participate in what the chemist calls tetrahedral bonding. The name refers to the fact that the angles of these four bonds are precisely those assumed by lines emanating from the center of a perfect tetrahedron outward to its four equally spaced vertices. In this tetrahedral bonding

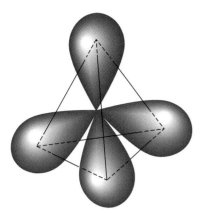 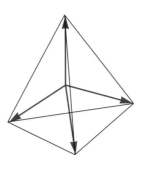

A column IVB atom such as silicon forms hybrid orbitals so that each of its four outer-shell electrons can bond covalently with neighboring atoms: the spherical *s* orbital and the three *p* orbitals, which originally extended along the x-, y-, and z-directions, merge, or "hybridize," to create this four-lobed configuration. The shape of the four hybridized orbitals is actually a bit more complex, with additional short opposing lobes opposite each larger lobe, but these details are left out for the sake of clarity. Shown alongside is a perfect tetrahedron, which defines the angles of this "tetrahedrally bonded" configuration.

configuration, each silicon atom arranges itself into the forming silicon crystal, shown in the figure to the right, in such a way that each of these four extended lobes overlaps with a lobe of another silicon atom. Each silicon atom thus shares an additional electron in each lobe, or four additional electrons in all. Note in the figure that each silicon atom bonds to precisely four neighbors. The bonds to atoms in the next cube of this repeating structure are shown extending slightly out of the cube to help the reader visualize the positions of the silicon atoms next to those lying on the edge of the cube. By gaining the four extra shared electrons, each silicon atom increases its electron count from 14 to 18, thus reaching the stable filled-shell configuration of argon. Diamond, a crystalline form of carbon (a Group IVB element like silicon), also forms in precisely this fourfold bonded structure, and it is one of the strongest materials known. Such is the strength of covalent bonding.

We've now viewed the formation of a solid, in this case silicon, from the chemist's point of view. We've watched the electronic orbitals of the isolated atoms of a gas begin to overlap to form chemical bonds as the atoms assemble into a solid. Although the more tightly bound electrons in the inner shells remain relatively unaffected by this process, the electrons in the unfilled outer shell are affected in substantial ways. The

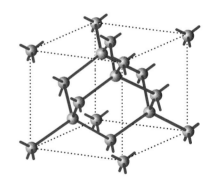

The crystal structure of silicon. Each atom shares the electrons of four neighboring atoms in four bonds, represented here by the red sticks. This arrangement brings the number of electrons surrounding each silicon atom up to the number in the "magic" filled 3*p* shell of argon. The dotted lines do not indicate bonds, but are added to aid the eye in visualizing the cubic cell of atoms, which is endlessly repeated to form the complete crystal.

shapes of their orbitals changes, and even the atom they are "assigned to" changes in a sense.

The energy levels most affected by the chemist's bonding process are precisely those important to the creation of laser light. Transitions between electronic states are the source of light in a gas laser, and the states participating most frequently in this process are those in the outermost shells, the part of the energy-level diagram where there is a chance of finding empty states into which excited electrons can fall. So in the process of forming a chemically bonded solid, we are significantly changing the energy-level scheme ultimately responsible for light emission. We shouldn't expect the energy transitions responsible for lasing to be the same in the solid as they are in the gas, and in fact they are not. These energy-level changes, touched upon earlier in Chapter 2, come into the forefront when we look at the formation of a silicon solid from a solid-state physicist's point of view.

The chemist's view of solids, insightful though it is, alone does not help us to distinguish between an insulator such as diamond and a semiconductor such as silicon. Position in the periodic table and crystal structure are important clues to an element's electrical properties but by themselves do not explain them. Physicists were thus stymied in their understanding of solids until the creation of the theory of quantum mechanics in the 1920s. By suggesting that each electron in a solid must have its own particular quantum state, the theory implied that many millions of energy states must exist in a solid with millions of atoms. This consequence stimulated the thinking of a number of physicists interested in the behavior of solids.

The Band in Solid–State Physics

In 1926 Erwin Schrödinger published his famous equation, a cornerstone of quantum mechanics, holding the key to the detailed electronic structure of matter of all types. His equation described the quantum mechanical behavior of an electron in any arbitrary situation in which the potential that affects its energy is known mathematically. Though solvable for the simple hydrogen atom, by the time the effects of the many atoms of a solid are entered into the Schrödinger equation, it becomes mathematically intractable. Nevertheless, the publication of Schrödinger's equation stimulated a burst of activity among physicists attempting to mine the insights hidden within it. One of these physicists was Felix Bloch, a young student in Leipzig, who in 1928 set out to calculate the molecular orbitals of a simple crystal. His mathematical approach, based on the fact that a crystal is a repeating structure in all directions, led to important simplifications that made it possible for the first time to begin to apply Schrödinger's equation to a solid. The discovery brought to light by his calculations, that atomic energy levels holding single electrons had broadened into energy bands holding many electrons, paved the way for the modern understanding of semiconductors.

Imagine a container of gas about the size of your fist. It would hold about 10^{22} identical atoms, each with a number of electron energy levels, or "states" in the parlance of the solid-state physicist, characteristic of that kind of atom, usually about ten or so. So within the gas there exist 10^{23} states in all. We push on the container's walls, squeezing the atoms within closer together. As we bring the atoms closer, individual atoms begin to interact with neighboring atoms and are brought toward the fixed positions of atoms in a crystal. Increasingly the atoms begin to interact with the rest of the new-forming crystal. The presence of nearby atoms and their surrounding electron clouds begin to perturb the energy levels of the atoms. This perturbation becomes greater and greater as the atoms are brought closer and closer together. By the time the atoms are only angstroms apart, the characteristic separation of atoms in a solid, they are no longer independent of the rest of the crystal at all. There still exist 10^{23} states, but now some of them are no longer assigned to an individual atom as they were in the case of the starting gas. Instead, 10^{22} or so different states are assigned to the crystal as a whole.

The sheer number of states suggests that they must be tightly packed in energy. Because there are so many, so closely spaced, they appear to merge to form continuous bands of energy as Bloch had shown in his elegant mathematics. This is precisely the process we discussed earlier in Chapter 2. This "band theory" approach is really just another way of looking at the sharing process that we observed earlier from the chemist's point of view. The bands that form in the solid-state physicist's model are equivalent to the bonding modifications that affect the atoms' electronic shells in the chemist's model. Whereas the chemist would say that all the silicon atoms have stable, filled-shell environments, the solid-state physicist would say that the band involved in the bonding (that is, the one resulting from the merging of all those levels in the original silicon atom that were occupied by the 4 extra electrons beyond the filled shell of 10) is in fact completely filled. This band theory of solids, fully developed by the mid-1930s, pointed physicists to the existence of semiconductors as a separate class of solids, along with conductors and insulators. The key to distinguishing the three classes of materials lay not only in the bands themselves but also in the gaps that appeared between these bands.

The same year that Bloch published his paper on energy bands, another young student of physics, Hans Bethe, made an additional fundamental contribution to the band theory of solids. While calculating what

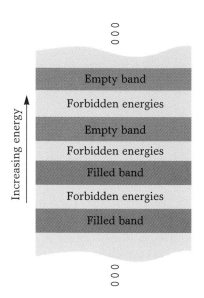

A solid-state physicist's view of the energy structure of a solid. As atoms are brought close together to form the solid, their uppermost filled energy levels broaden into filled bands, shown here in red, and the unfilled levels above them broaden into empty bands, shown in blue. The key to the properties of a semiconductor lies in the energy gaps—the forbidden energies shown in yellow—that separate these various bands.

happens as an electron travels within a crystal, he found that ranges of energies develop at which no electron can exist. These ranges correspond precisely to the gaps between the bands that Bloch had already noted. Both Bethe's and Bloch's work focused subsequent research in the field onto the question of how these gaps affected the crystal's electrical conductivity and other properties.

The concept of an energy gap should be familiar from Chapter 2, where we noted that the energy structure of a solid is characterized by a series of energy bands separated by gaps. The lower bands are filled from the bottom up, in the same way a bathtub fills, until all the available electrons have found states in the lower-lying bands. In that discussion, we used this model to explain the physical origin of transparency—why some solids fail completely to absorb some colors of light.

Conductors, Insulators, and Semiconductors

A number of the important properties of a given type of solid depend on precisely where the "waterline" in the bathtub model ends up. When all the electrons in the solid have been poured into the specific energy-band structure of a particular solid, where is the top of the filled portion? If it is somewhere in the middle of an allowed energy band, as shown in the left panel of the figure on the following page, then any amount of energy imparted to the electrons sitting at the very top of the filled portion of the band will promote those electrons into an allowed empty state in the unfilled part of the band immediately above. Even the small amount of energy associated with the ambient temperature is enough to do this. As a result, these promoted electrons, which prior to their promotion were laterally hemmed in by adjacent filled electron states and hence unable to move, are now free to move either to the left or right in the figure, corresponding to movement in the solid.

A metal is an example of such a solid, and many of the physical properties we have come to associate with metals are a result of this physical fact. Metals are quite good electrical conductors, for example, because of the freedom of motion given to their thermally excited electrons. In metals high thermal conductivity is also a direct macroscopic manifestation of this microscopic electronic property, for the freely moving electrons carry the extra heat energy from the hot side of a crystal to the cold side. And finally, remembering our previous discussion of transparency as a result of a finite forbidden energy gap, the fact that metals are rarely transparent comes from the fact that there is no forbidden gap between the filled and empty states. Hence they can absorb most energies of light.

A contrasting class of solids consists of those in which the dividing line between the filled and empty states is in the energy gap between

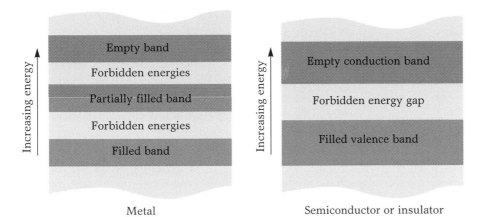

Metal   Semiconductor or insulator

The band structures of the three major classes of solids. The uppermost electrons (shown in orange) in a metal, depicted on the left, only partially fill a complete band. As a result, even a small amount of energy can promote electrons lying right at the top of the filled portion of that half-filled band into an empty blue area. Semiconductors or insulators, depicted on the right, have an uppermost occupied band that is completely filled. Their electrons are hemmed in on all sides by other electrons and are unable to move.

two allowed bands, in other words within the yellow forbidden energy gap in the right panel of the figure above. In these solids, the electrons in the entirely filled orange bands are packed in so densely that they are unable to move. And in the higher-lying blue bands, where there are a plethora of empty states in which to move, there are no electrons! Such a solid has no mobile electrons and is unable to conduct electricity at all. Both insulators and semiconductors have just such a band structure. The distinction between the two has to do with the size of the forbidden energy gap that separates the highest filled band from the lowest unfilled band.

We mentioned above that, in the case of metals, even the slightest amount of excitation energy imparted to the electrons at the top of the half-filled band will promote them into the unfilled part of the band where they can move. The thermal energy of the crystal itself is sufficient to promote electrons into the unfilled, upper portion of the band of a metal. This source of energy is also available in insulators and semiconductors, since they too have a finite temperature, but here the electrons have to cross a forbidden energy gap immediately above the highest filled electron states. Since electrons are unable, by quantum mechanics, to occupy these forbidden states, they must receive a certain *minimum* amount of energy equal to the width of the energy gap of the material (the material's "bandgap" or width of the yellow band) before they can be promoted at all.

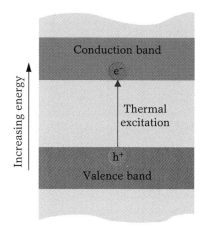

Because a semiconductor has a relatively narrow energy gap in comparison to an insulator, thermal excitation can promote electrons (e$^-$) from the filled valence band into the empty conduction band where they are free to move. The excitation across the energy gap leaves a hole (h$^+$) in the valence band in the process.

What is important, therefore, is the size of the bandgap in comparison with the typical energy available from the heat in the crystal. This comparison provides the key difference between insulators and semiconductors. The forbidden energy gap of an insulator is so much larger than the average thermal energy that virtually no electrons are thermally promoted to the upper unfilled band. From the *lack* of mobile electrons in the upper band, referred to as the conduction band since it is there that the conduction would take place, the insulator derives its key property of being a poor conductor of electricity.

On the other hand, the bandgap in a semiconductor is narrow enough that a small number of electrons from the top of the highest filled band (referred to by solid-state physicists as the valence band) are able to be promoted into the bottom of the empty conduction band solely from the thermal energy available in the crystal. These promoted electrons in the conduction band are then free to travel laterally, and hence the semiconductor is able to conduct electricity. Because only a few electrons are excited across the finite energy gap, however, the number of thermally promoted electrons in the conduction band is very limited. As a result, the number of free electron carriers is typically more than a billion times smaller in a semiconductor than in a metal. So semiconductors conduct electricity, but not nearly as well as true conductors like metals. Hence the name "semiconductor."

The band theory of semiconductors explains many of the curious properties noted in these materials beginning with Faraday's 1833 observations. A semiconductor's electrical conductivity rises with temperature, for example, because a greater number of electrons are promoted into the conduction band in the presence of more thermal energy. The increase in conductivity upon illumination can also be understood as the result of an increased number of conduction band electrons, created, in this instance, as the absorption of light provides the energy to promote electrons across the energy gap.

If this were the whole story on semiconductors, they would be relegated "to the bench" as a poor second-string substitute for a good conductor. They do possess the interesting property that as we add more thermal energy (i.e., heat them up), the number of conducting carriers increases, resulting in a conductivity that is distinctly temperature dependent. This property can be exploited in some narrow classes of devices such as thermistors, whose resistance is a well-characterized function of their temperature, making them useful as the temperature-measuring device in inexpensive digital thermometers. But if that were the only property of semiconductors of technological interest, we probably never would have heard of them. Yet by some people's accounts, we live today in the Silicon Age, the Age of the Semiconductor, so there must be something we are not capturing in this simple description. And in fact there is.

Semiconductors and Doping

There is a second sense in which these materials are "semiconducting," arising from the way their electrical properties change radically by the addition of small amounts of controlled impurities. If the impurity is an atom from the same column of the periodic table, like germanium when added to silicon, the electrical properties remain unchanged. (There is a subtle effect of changing the bandgap, but we'll ignore that here and return to such effects in the next chapter.) The dramatic changes come when we add an impurity from a different column of the periodic table. Such an impurity has a different number of electrons to contribute to the silicon solid than does the silicon atom it replaces, which contributes four electrons from each atom.

Let's look first at an impurity, such as arsenic (As), from column VB. The elements in this column have five electrons in the outer shell, rather than the four of silicon. Suppose we replace a silicon atom in our silicon crystal with an arsenic atom. Four of the arsenic's five electrons each participate in one of the four bonds joining the arsenic atom to the neighboring silicon atoms, as was the case for the silicon atom it replaced. But that leaves one *extra* electron. Where does it go? There is no room for it in the filled valence band, so it is donated to the empty

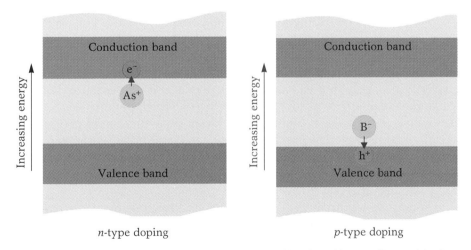

n-type doping   *p*-type doping

The conductivity of semiconductors can be altered by the addition of impurities in the process of "doping." In the left-hand panel, an arsenic (As) atom is substituted for a silicon atom in the crystal lattice, and it donates its extra electron to the bottom of the conduction band, the lowest available place to put it. In the right-hand panel, a boron (B) atom replacing a silicon atom must steal an electron from the filled valence band, leaving a hole, in order to have enough electrons to bond with its four nearest neighbors in the lattice.

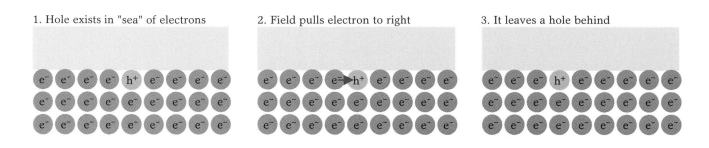

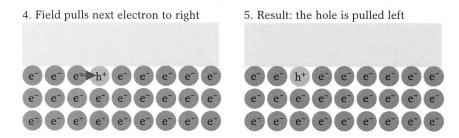

An electric field pulling electrons to the right has the net effect of making a hole in the nearly filled valence band move to the left.

conduction band, as illustrated in the left panel of the figure on the previous page. This type of impurity is therefore called a "donor."

This process of adding donor atoms is called doping. For every arsenic impurity we add to the silicon, we get one more electron available for conduction. And unlike the thermal case, where the number of carriers varies dramatically with temperature, we can now control the conductivity precisely in a way almost independent of temperature. We can make the conductivity as low as that of the undoped, or "intrinsic," semiconductor, more than a billion times lower than the conductivity of a metal, or as high as that of a reasonably good conductor, perhaps only a hundred times lower than that of a metal. The conductivity of the semiconductor is limited only by how many atoms of the impurity we can force the silicon crystal to accept.

Adding an impurity with extra electrons thus transforms the crystal into an electrical conductor. An analogous scenario develops when an impurity with one less electron (from column IIIB of the periodic table) is added. Atoms of the element boron, for example, serve as such an impurity known as an "acceptor." Starting with only three electrons in its outermost shell, a boron atom needs to rob the crystal of an electron in order to fulfill its need for four electrons to bond with its four nearest silicon neighbors. It does this by accepting an electron from the sea of electrons in the filled orange valence band, as shown in the right panel of the figure on page 83. Now instead of an extra electron in the conduction band, we have an empty space or hole in the previously filled valence band.

These empty holes have the same ability to transform the silicon into an electrical conductor as the extra electrons of arsenic do. This property may seem puzzling at first, but it can be illustrated by examining how the configuration reacts to an external electric field. If we apply a field that causes electrons to be pulled to the right, then the electron to the immediate left of our hole will jump to the right, filling the hole. In the process, this creates a hole in the electron sea at the place from which the electron jumped, the space immediately to the left of where the original hole was. This hole then acts as a space into which the electron to *its* immediate left jumps, thus moving the hole one more hop to the left. Thus, the effect is a steady movement of the hole in the left direction. Although the physical process involves the movement of real particles, electrons, to the right, the net effect is the movement of a pseudoparticle, a hole, to the left. In other words, the hole acts like a positively charged particle in that in the presence of an electric field it moves in the opposite direction to that of a negatively charged electron.

So the net effect of adding impurities such as boron to silicon is to add positively charged holes to the crystal. Like the electrons donated by the arsenic, the presence of the holes can change the conductivity of the crystal over many orders of magnitude. Thus with the addition of donor or acceptor impurities (like arsenic and boron), we gain the powerful ability to tailor the conductivity of the crystal to almost any value we choose. In the next chapter, we will see how layers of donor-doped and acceptor-doped semiconductors, when placed directly adjacent to one another, can be the source of light we need to construct an all-solid-state semiconductor laser with dimensions measured in fractions of a millimeter.

Tiny fiber optic strands made of ultrapure glass are capable
of guiding communications signals encoded as pulses of light
from advanced semiconductor lasers. Fiber strands can
transmit billions of bits of information a second over
distances of many kilometers.

5

The Semiconductor Laser

It wasn't long after the invention of the gas laser—a bare two years—before researchers realized that, if semiconductors could be made to release photons of light, they could be used as the basis of lasers. These new semiconductors would not merely duplicate the functions of gas lasers, however, but would find their own unique uses. While gas lasers were proving their use in laboratory research and heavy duty industrial applications, a few far-sighted individuals were envisioning an entirely new application for lasers—these devices could become the heart of a new form of communication. Data, telephone conversations, and TV signals could be sent as pulses of light instead of electric currents, achieving communication at the speed of light. Today this vision is coming to fruition with the spread of fiber optic systems. Lasers are needed to generate the optical signals that travel through these systems, but gas lasers cannot do the job in a practical manner.

Gas lasers range in length from a few inches to tens of feet; the light beams they produce measure as low as milliwatts in power and as high as many megawatts. For the networking of optical communications systems, however, even the smallest of gas lasers is much too large to be practical.

A single optical fiber is about the size of a human hair. Its outside diameter of about 120 micrometers (or millionths of a meter) is only slightly larger than that of a hair, which has a thickness in the range of about 50 to 100 micrometers. But whereas human hair is at most a few feet in length, fiber optic strands can stretch for many kilometers. Made of quartz, these strands are so clear, and carry light with such low losses, that laser light sent through fiber over a distance of about 200 kilometers can still be detected at the far end. Fiber optic systems are thus very practical for communications across the ocean or for any long-distance terrestrial connection. Because of their compact size, light weight, and secure transmission properties, they are also ideal for short distance communication systems, as well.

An optical fiber, shown schematically in the figure on this page, has a central core about 7 micrometers in diameter through which the light travels. The type of fiber illustrated, called a single-mode fiber, is the most common type used for fiber optic communications. A light beam must be injected into the end of such a fiber in such a way that it can be trapped and guided in the narrow 7-micrometer core. A miniature laser about the size of a grain of salt, or smaller, is just right for this purpose. And this is about the size of modern lasers made from semiconductor materials.

Having seen the rise of the Electronic, or Semiconductor, Age, we are now seeing the beginning of the Age of Optoelectronics. Some of the most popular of modern electronic equipment combines integrated circuits with optical devices such as lasers. Compact disc players are a common example; a semiconductor laser combined with silicon integrated-circuit electronics now provides the signal processing for sound production in home stereo systems. Optoelectronic technology is also used to access the information on video and CD-ROM discs. The integrated optoelectronic circuits of these devices achieve fast, sophisticated signal processing.

A gas laser must have a quartz tube to confine the gas and a pair of precisely aligned mirrors to form the laser beam. The beauty of the solid-state laser is that the semiconductor material itself does both these

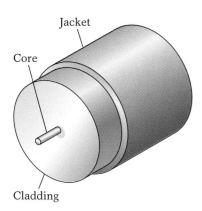

Jacket

Core

Cladding

A cross section of a single mode optical fiber, barely larger in diameter than the thickness of a human hair, used to guide the pulses of light from a semiconductor laser. The ultrathin core, made of extremely clear doped quartz and only seven millionths of a meter in diameter, has a higher index of refraction than the surrounding quartz cladding in order to guide the light down this central axis of the fiber. The plastic jacket protects the inner glass from the deleterious effects of the surrounding environment.

things. The economy of the arrangement accounts in part for the laser's miniature size.

The figure on this page shows a schematic of a typical semiconductor laser. First notice that it consists of a set of very thin layers of different semiconductor materials stacked on a thicker layer of semiconductor material, the "substrate." Virtually all optoelectronic devices are made of thin layers of dissimilar materials. In a typical laser, all the layers together are less than 10 micrometers in thickness, while the substrate has a thickness of a few hundred micrometers. Stable bonding of atoms in the solid material takes care of holding the atoms together, replacing the quartz tube used for containment in the case of the gas laser. Reflecting mirrors are easy to obtain: since semiconductors are crystals, they can be cleaved so that the crystal facets serve as mirrors. In the figure, the front and back faces are cleaved facets forming the laser optical cavity.

In the schematic, laser light emerges from the structure at the spot marked in red on the cleaved facet. Light is generated in one particular layer of material, called the active layer. This very thin layer is only about 50 to 1000 atoms thick. Physicists of the early 1970s first figured out how to coax light from such a thin layer of material. Their earliest efforts were considerably simpler in structure than the multilayered laser shown to the right. In fact, the first semiconductor lasers consisted of only two layers. To understand how these devices worked, we take a closer look inside the active layer.

Light from a Semiconductor

We've seen that semiconductor crystals can be "doped" with two different kinds of impurities, those that add extra electrons to the crystal and those that add extra "holes." Semiconductors doped with electron-adding impurities are referred to as n-type, because the extra carriers being added are negatively charged, whereas those doped with hole-adding impurities are called p-type, because the carriers are positively charged. Increasing the amount of dopant in the crystal increases the electrical conductivity of the material, regardless of whether the dopant chosen is n- or p-type. So why does the distinction between the two matter? The answer was first glimpsed around 1940 by scientists at Bell Laboratories.

Two metallurgists at Bell Laboratories, J. H. Scaff and H. C. Theuerer, had discovered that by melting and resolidifying silicon in a vacuum, they could obtain relatively pure ingots. Yet some slight natural contamination was inevitable: they noticed that silicon contaminated with phosphorus conducted best when a negative voltage was applied (n-type silicon), whereas silicon contaminated with boron or indium conducted best

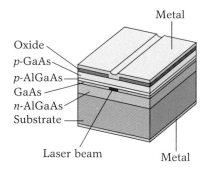

A semiconductor laser is built up of a series of layers of different materials, including, in this case, gallium arsenide (GaAs) and aluminum gallium arsenide (AlGaAs). The n- and p- prefixes refer to the type of dopant the layers contain. The figure is not drawn to scale, since the substrate on which the complex layer structure is deposited is a few hundred micrometers in thickness, while the entire set of layers above it is less than 10. It is within this set of layers that the current, which flows into and out of the device through the metal layers at the top and bottom, is converted to light. Confined to the thin central GaAs active layer, the light emerges from the front flat crystal facet from the region colored orange.

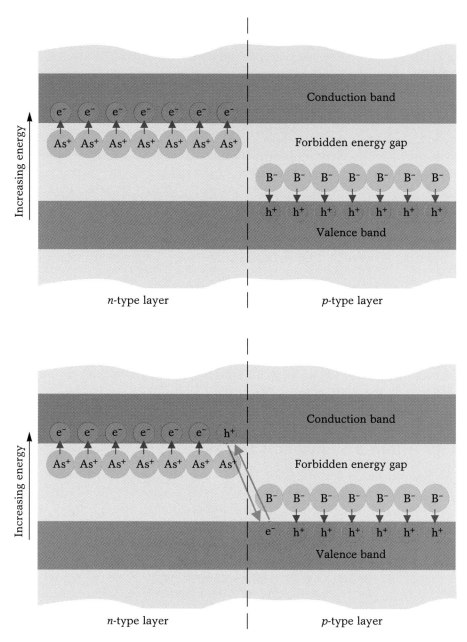

This example of a two-layer structure known as a "*pn* junction" consists of a silicon semiconductor layer doped with an *n*-type dopant, arsenic (As), on the left immediately adjacent to a similar silicon layer doped with a *p*-type dopant, boron (B), on the right. In the lower panel, the rightmost electron in the left-hand *n*-type layer moves right into the *p*-type layer, where it drops down in energy by falling into the closest hole, thus annihilating it. The movement eliminates both an electron and a hole near the interface.

under a positive voltage (p-type silicon). In 1940, a chemist at the laboratories, Russell S. Ohl, began to experiment on an ingot with an especially sharp boundary between a p-type region at one end and an n-type region at the other. When Ohl shined a light directly at the boundary between the two regions, he was amazed to see a voltmeter connected to the silicon deflect by half a volt, ten times more than anybody at the time would have expected. This was not the first time that semiconductors had been observed to convert light to electrical energy, but this time the reading was 10 times larger than any Ohl or his colleagues had seen before.

Ohl's experiment set other scientists at Bell Labs thinking; eventually their work led to the creation of the first transistor, in 1947. Like Ohl's silicon ingot, transistors consist of regions of p-type and n-type semiconductors. With their ability to amplify electric currents under controlled conditions, these devices can perform fast on-off switching that is essential to the operation of modern computers. After the invention of the ruby crystal and gas lasers, the success of transistors inspired some researchers to look at semiconductors as a laser medium.

Suppose we fabricate a structure that consists of two different layers. Both layers we make from the same semiconductor material, say silicon. The only difference in the two layers is the type of dopant we add. To the left-hand layer, we add an n-type dopant, which leaves it with an excess of electrons in the mostly empty conduction band, as shown in the upper panel of the figure on the facing page. To the right-hand layer, we add a p-type dopant, which leaves it with a number of holes in the mostly filled valence band. The resulting structure is called a pn junction, which is of fundamental importance in semiconductor electronic and optical devices.

The upper panel of the diagram indicates the state of affairs right at the moment that we bring the two layers into contact with each other. If we look at it closely, we see that this situation is unstable. Consider the rightmost electron in the n-type layer. Within its own layer, it has nowhere lower in energy into which it can spontaneously drop, because the valence band beneath it is entirely filled with electrons (solid orange). The situation changes, however, when we allow this n-type layer to come into contact with the adjacent p-type layer. Now, immediately to its right in the most left-hand portion of the p-type layer, there exist holes in the valence band, caused by the p-type dopant (boron in this case) snapping up valence band electrons. That rightmost electron in the n-type layer's conduction band quite happily moves over slightly to the right, drops appreciably lower in energy, and fills the leftmost hole in the p-type layer, as indicated by the orange arrow in the lower panel of the figure on the preceding page.

What Is a "Hole"?

Holes can be thought of as a place in the filled valence band where there are no electrons, much as one can think of a bubble in a carbonated glass of soda as a place in the soda where there is no liquid. We can consider the movement of bubbles to the top of the glass in two ways. We start with the fact that the liquid that makes up the soda itself is denser than the gas in the bubbles. The bubbles move upward because gravity pulls more strongly on the dense liquid above the bubble, causing that liquid to move down into the point in space occupied by the bubble below. In the process, the gas of the bubble is displaced into the space previously occupied by the adjacent denser liquid, and the result is that the bubble (or "lack of liquid" in a sense) moves upward. In flowing downward, the liquid system naturally lowers its energy.

That's a rather complex way of looking at the movement of bubbles, although it gets back to the force at the root of the phenomenon, namely that of gravity on the liquid. A much simpler way of thinking about the ascension of bubbles is to construct a force called buoyancy that is directed upward on objects less dense than the surrounding fluid. The phenomenon of buoyancy of

Bubbles rising to the top of a carbonated drink serve as an analogy for "holes" in the filled sea of electrons. Just as the force of gravity pulling the liquid down results in the bubbles rising up, so the tendency of electrons to fall to lower energy states results effectively in the tendency of holes to rise up in energy.

course owes its existence to gravity pulling the adjacent denser material downward in precisely the way just described. But when we wrap that

One can just as easily look at the process from the hole's point of view. A hole attains a lower energy not by dropping *down* in the diagram like an electron does in losing energy, but rather by floating *upward,* in a way entirely analogous to a less dense gas bubble in a liquid being forced upward by the downward pull of gravity. The hole in the figure on page 90 achieves this lowering of energy by sliding over a hair to the left into the *n*-type layer and then floating up to replace the rightmost electron in the conduction band, as indicated by the blue arrow. Whether one views the process from the point of view of the hole or the rather complex sequence of events into the "constructed" force of buoyancy, we get a much simpler description of what happens to the bubble. The bubble is less dense than the liquid. Hence the buoyancy force pulls it upward. Simple.

Constructing a new concept for the sake of simplification is precisely what solid-state physicists do in their description of holes in semiconductors. Like the bubble in the soda, a hole will naturally tend to move upward in energy diagrams like those on pages 90 and 94 due to the fact that the electrons which form the "sea" around it tend to drop downward to lower energy. In a similar fashion, the process of a hole moving in response to an externally applied electric field, described in the figure on page 84, is completely analogous to the upward movement of soda bubbles. The negatively charged electrons are the real particles being pulled by the electric field; in that figure they are moved to the right. The holes, or "lack of electrons" if you like, aren't directly pulled by anything. They're not even particles; they're a "lack" of a particle. But the net effect of the electric field pulling the real electrons to the right results in the movement of the holes to the left. The holes act as though they are real particles reacting to a force that is pulling them in a direction opposite to that exerted by the electric field on the electrons, just as the bubbles act as though they were a real "liquid" with a buoyancy force pulling them in a direction opposite to that of gravity.

In fact, the holes behave exactly as real positive particles would behave if any were present. The nature of an electric field is that it pulls charges of opposite signs in opposite directions. So our "constructed" force pulls the holes in the same direction that an electric field would pull a particle of charge opposite to that of the negatively charged electron—namely, a positively charged particle. This is not surprising since the lack of an electron in a sea of negatively charged electrons is in fact a local area of relatively more positive charge.

Even though holes are a "lack of electrons," we find it easier to consider them to be simple, positively charged "particles" that, because of their charge, are pulled in a direction opposite to that of an electron in the presence of an electric field. This is the model we will adopt for the remainder of our discussion, although if the reader prefers to think of a hole as a "lack of electron," all the ensuing discussion can in principle be cast into that framework. The mental gymnastics, however, might prove to be rather complex.

electron is like the difference between describing the glass as half full or half empty. The result is the same.

There are some strange conceptual issues that arise because of these two possible viewpoints. We have no problem thinking of an electron filling in or replacing a hole, yet we find it strange to think of a hole replacing an electron. But this is precisely what we must think: the hole moves from its position in the valence band, leaving behind an electron (the absence of a hole), and goes into the adjacent conduction band, where it replaces an electron.

Top: The charged As$^+$ and B$^-$ left behind in the transfer of an electron to the right make the transfer of the next electron to the right less energetically favorable than the first. This retarding effect is depicted in the diagram as a relative lowering of the entire set of energy bands in the *n*-type material on the left-hand side of the figure. Bottom: The transfer of a second electron to the right exposes more charge, further dropping the energy bands of the left-hand *n*-type layer, and thus further increasing the retarding effect to be overcome by the next electron to travel into the *p*-type layer.

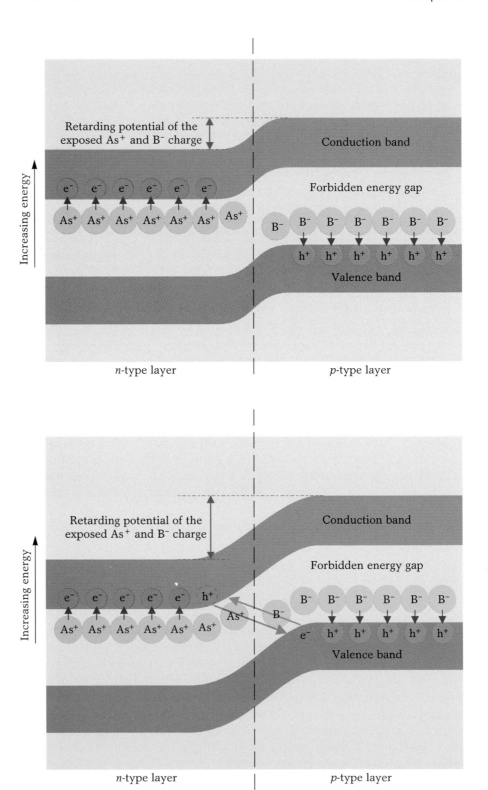

Now if one electron can move across to the right and drop in energy into a hole, why can't all the electrons do so? The answer becomes clear when we consider the cumulative effect of any electrons that have previously moved to the right. The reader will notice in the bottom panel of the figure on page 90 that, although an electron-hole pair has been annihilated, the exposed dopant atoms from which that electron and that hole arose remain in place (the As^+ and B^-). And those atoms still possess electric charges: positive on the arsenic atom and negative on the boron atom. The positive arsenic atom attracts the negative electrons residing nearby in the conduction band and thereby partially holds back any additional electrons that might otherwise move into the hole states of the p-type layer. In an identical manner, the negatively charged boron atom repels those same electrons, further discouraging their lateral transfer. The collective effect of such a retarding field, holding the electron back from transferring, can be represented in the figure as a shift in the heights of the energy levels in the two layers. The amount of the shift represents the retarding effect of the exposed As^+ and B^- charge. Lowering the electrons' energy on the left-hand side accomplishes the task of reducing the driving potential for electrons and holes to move. This energy change is represented in the figure on the opposite page as the amount the two sides are shifted relative to each other, as indicated by the vertical blue arrow. The energy change, of course, depends in turn on the amount of exposed As^+ and B^- charge.

In the upper panel of the figure, there is still a "hill" for electrons to drop down and reduce their energy, so the next electron lowers its energy by moving across the junction between the two layers, as depicted in the lower panel. The result, however, is to further increase the amount of charge from the now newly exposed As^+ and B^-, so the relative energy shift in levels between the two sides increases. This process continues until a point of equilibrium is reached: at this point, the steadily increasing energy shift created by the increasing amount of exposed charge of the dopant atoms exactly balances the total amount of energy gained by electrons falling across the interface from the n-type into the p-type material. This equilibrium state is shown in the figure on the next page. The energy shift, represented by the length of the vertical blue arrow, becomes as large as the size of the energy gap, and the electrons in the n-type material and the holes in the p-type material are at approximately the same energy level. Therefore there is no further net transfer of charge across the interface between the two layers of the pn junction, and an equilibrium is established. This structure, referred to as a "diode," is basic to all semiconductor electronics and optoelectronics.

So how is all of this used to obtain light for our semiconductor laser? As noted in the figure on page 90, the electrons moving to the p-type

At equilibrium, the energy level of both the electrons and holes becomes comparable, no more net charge travels across the junction interface, and a steady state is achieved. Such a structure will conduct electricity if a small voltage is applied which counteracts the retarding potential, thus flattening the bands, but will not conduct for voltages applied in the opposite polarity. Devices with asymmetric conductivity such as this are called diodes.

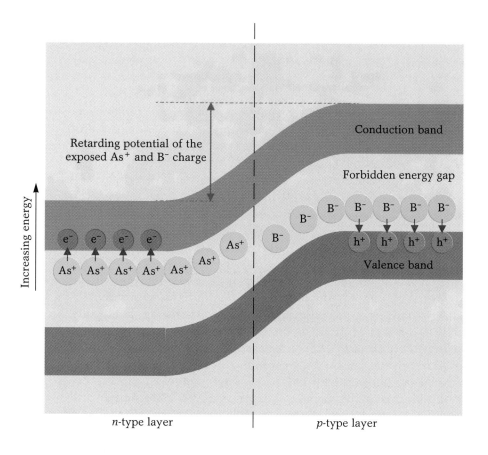

n-type layer | *p*-type layer

layer drop in energy an amount equal to the forbidden energy gap of the semiconductor. The point we have thus far failed to discuss is what happens to the energy that is released when the electron drops down. Because of the law of conservation of energy, this small chunk or "quantum" of energy cannot disappear altogether. Once it is lost by the electron it must appear in some other form within the crystal.

The form in which it appears depends on the type of semiconductor crystal. In the case of the most widely used semiconductor, silicon, the energy is distributed into the crystal in the form of heat. This production of heat energy causes the individual atoms of the crystal to vibrate a little more, thus slightly increasing the crystal's temperature. Such an outcome is not very exciting, and in fact the dissipation of heat in a semiconductor is more of a nuisance than anything else. One must mount devices that generate a large amount of such heat on large metal heat sinks to carry the heat away before the devices become too hot. The silicon chip serving as the central processing unit of a modern personal computer generates so much heat, for this reason and others, that it needs a small fan attached to its top, running the entire time the machine is on, just to keep it cool!

But there exists a certain class of semiconductor in which the energy released by the drop of an electron across the forbidden energy gap can be released not as heat but as a quantum of light, just like the emission from atoms discussed earlier in Chapter 2. These semiconductors fall into a general category called "compound semiconductors" because they are formed as compounds of two or more elements from the periodic table. Examples are the compound semiconductors gallium arsenide (GaAs), aluminum gallium arsenide (AlGaAs), and indium gallium arsenide phosphide (InGaAsP). If we locate the atoms from which these compounds are constructed in the periodic table, we find that they come from columns IIIB and VB. Hence these materials are designated III-V semiconductors (pronounced "three-five"). They form crystals in precisely the same geometric arrangement as silicon, shown in the figure on page 77, but with the difference that half the atom positions are occupied by Group IIIB atoms (the gallium, indium, aluminum, etc.) and the other half by Group VB atoms (such as arsenic and phosphorus).

The "fifty-fifty" mix of the Group IIIB and Group VB atoms is necessary to fill the proper number of electrons into each of the four bonds attached to each atom. In pure silicon, each silicon atom contributes one of its four available bonding electrons to each of the four bonds. Everything comes out nicely. In a III-V semiconductor, the Group III atom has only three available electrons. But to form the structure shown in the figure on this page, it needs four. The extra electron comes from an adjacent Group V atom, whose five available electrons are one more than is needed for its four bonds. As long as there are equal numbers of Group III and Group V atoms in the structure, the number of electrons comes out correctly. In fact, the same procedure also can be carried out one more column away in the periodic table by combining equal parts of Group IIB and Group VIB atoms. Each Group VI atom contributes its *two* extra electrons to its neighboring two-electron-deficient Group II atom. Again, the total number of Group II atoms must be equal to the total number of Group VI atoms. Examples of such II–VI compounds are mercury cadmium telluride (HgCdTe), used as the sensing element in night vision goggles, and zinc selenide (ZnSe), one of the basic materials in the new blue-light semiconductor lasers being designed for the latest generation of ultra-high-density optical disc storage devices and color displays.

The details of the complex energy structure of these materials that result in this ability to produce light are beyond the scope of this book.

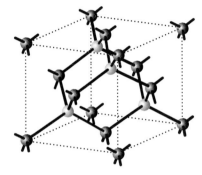

● Group III atom such as Ga

◯ Group V atom such as As

The crystal structure of a compound semiconductor such as gallium arsenide (GaAs) is similar to the crystal structure of silicon shown on page 77, with the difference that half the sites are assigned to group IIIB atoms (shown in purple) and half to group VB atoms (shown in green). Each bond joins a IIIB and VB atom, so that the electron deficiency of the group IIIB atom (three electrons available for four bonds) is made up by the excess electron on the VB atom (five available electrons for four bonds.)

Suffice it to say that, depending on the band structure, some semiconductor materials have this important property that electrons can drop directly across the energy gap in a single jump, producing a photon of light. These semiconductors are referred to as "direct gap" materials, and they provide the basic building blocks from which are constructed all semiconductor lasers and many light-emitting diodes (LEDs). Though devices based on these direct gap materials represent only a small fraction of the total number of semiconductor devices manufactured, the fact that the workhorse semiconductor silicon is an indirect gap material dictates the use of these direct gap compound semiconductors in all optoelectronic devices such as lasers.

The Earliest Semiconductor Lasers

The first semiconductor lasers, reported as early as 1962, were made from the material gallium arsenide. The inventors of the laser started with a substrate of *n*-type gallium arsenide, made by including an impurity such as selenium on an arsenic site that provides an extra electron not needed for the usual gallium-to-arsenic bonding. On top of this *n*-type layer, the laser designers deposited a layer of *p*-type gallium arsenide, made by adding an impurity such as zinc whose atoms are one electron short of the number in the gallium atoms they replace. The two layers together produced a wafer with a total thickness of around 500–750 micrometers. Next, the designers pared the wafer down until it had, typically, a length of about 300 micrometers and a width of about 200 micrometers. The resulting gallium arsenide device, a simple *pn* junction, had the potential to generate photons as the electrons drop across the central band gap. But, as we noted earlier, this junction between the *n*-type and *p*-type material very quickly reaches the equilibrium state. No further light can emanate from such a structure.

Fortunately, there is a simple way to create a continuous beam of light: we apply an electrical voltage to the *pn* junction depicted in the figure on page 96, by, for example, connecting a battery to it. By attaching the battery in the polarity shown in the figure on the facing page, we cause the bands on the left-hand side to be raised in energy relative to those on the right by an amount proportional to the battery's voltage. In the case depicted in the figure, that voltage is chosen to be just enough to boost the bands until they are back to being flat relative to each other. The positive side of the battery, connected to the right-hand portion of the structure, attracts electrons, which are of the opposite charge, out of the *p*-type material. The exiting electrons create a continuous supply of holes in that material. Within the battery, the electrons are boosted in

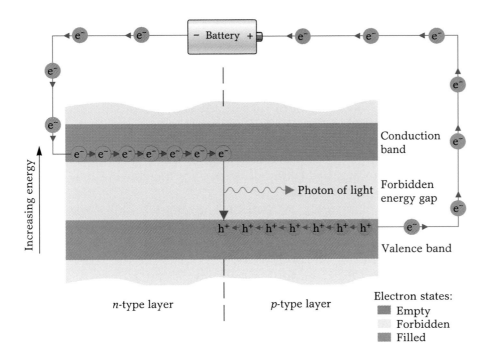

Increasing energy →

- Battery +

Conduction band

Photon of light — Forbidden energy gap

Valence band

n-type layer

p-type layer

Electron states:
■ Empty
□ Forbidden
■ Filled

An external electric voltage, applied here by means of a battery, can overcome the retarding effect of the exposed charge at the junction. The left-hand n-type layer is brought back up in energy relative to the right-hand p-type layer due to the effect of the battery, causing the bands at the interface to once again become flat with respect to each other. A continuous stream of electrons and holes flows into the central interface region, resulting in the continuous creation of photons in that region. This process serves as the source of light for a number of devices, including the light-emitting diode (LED) and the semiconductor laser.

energy so that as they enter the n-type material from the left, they are at the bottom of that material's conduction band.

Now when the electron shown poised in the center drops across the gap to release a photon, there is a continuous supply of new electrons coming from the negative side of the battery into the n-type material from the left. These electrons flow toward the center, and one takes the place of the electron that has dropped into the valence band. At the same time, the continuous drain of electrons out of the p-type material on the right creates a continuous source of valence band holes. These are free to flow leftward to the center to provide new holes into which the electrons from the n-type material drop.

The result is a steady state in which, through the wires, the battery continuously extracts electrons from the p-doped side of the device, and delivers them to the n-doped side of the device, where they are released at a higher energy (equal to the battery's voltage). To complete the circuit, the electrons driven across the interface from the n-doped side to the p-doped side fall one after the other across the bandgap, producing a continuous source of photons. We have created a device that converts electric current, the flow of electrons, into a stream of photons, the light that will power our semiconductor laser.

What we have in the figure on the preceding page is not a laser yet. This device is a light-emitting diode, or LED, and forms the basis of the red and green glowing numerals and blinking lights found universally in the displays seen in all sorts of electronic devices. Because these devices are so efficient at producing light, they are even replacing conventional lights in applications such as the red taillights in the newest generation of automobiles. The LED is the semiconductor equivalent of the glowing neon sign with which we began our discussion of the gas laser in Chapter 3. As with the neon sign, to make such a light source into a laser one requires a way of arranging this glowing light into a line between a pair of mirrors.

In the gas laser, the glowing light is arranged within a glass tube along a thin line passing through an optical cavity formed by placing mirrors at either end. The light amplification process continues only for photons traveling along this particular path, since they are the ones that are reflected back and forth many times by the mirrors. It is actually easier to form such a cavity in the case of the semiconductor laser than in the case of the gas laser. Taking advantage of the crystal properties of semiconductor materials, the designers of such lasers break by cleaving two parallel facets to form the reflecting mirrors. Such facet mirrors reflect about 30 percent of the light reaching them, a more than adequate magnitude for laser action because of the high efficiency of light emission in semiconductors. The other 70 percent is transmitted out into the air. The side faces of these lasers need to be rough compared to the cleaved ends so that the optical cavity is well defined for one direction only. These side faces are usually not formed by cleaving, but instead are scored and broken, as one cuts panes of glass, or else sawed using a high-quality circular diamond saw.

To complete the fabrication of their laser, the designers of the early 1960s had to add metal to the top and bottom surfaces of the laser device so that electric current could be applied. The laser was then bonded down to a metal plate, using a metal such as indium as a kind of conducting glue. A springy metal tip made from a material such as phosphorbronze was put in contact with the top of the device. After suitable connection to a source of electricity, pulses of current were passed through the device from top to bottom. The electrons flowed across the interface from the n-type into the p-type material, producing a large density of electrons in the conduction band. These electrons, now in the p-doped side, were thus mixed among a large density of holes in the valence band. It is these electrons that dropped across the bandgap to produce photons.

As in the neon laser, the first few photons created in this way are produced by spontaneous processes. Very soon, however, one of these early photons finds itself traveling along parallel to the junction interface and at the same time perpendicular to the two cleaved end facets. In this region it encounters many electron-hole pairs just waiting to combine. As this photon passes near an electron in the conduction band, it can stimulate the electron to fall into one of the holes in the valence band, producing a second photon identical to the first, moving parallel to the interface and aimed at one of the mirror facets. These photons are reflected from the mirrors and travel back and forth perpendicular to the mirror facets, creating the buildup of light that forms the laser beam. In this *pn*-junction laser device, sometimes referred to as a semiconductor diode laser, the light emerges from both cleaved mirror facets.

The buildup of light can happen as long as there is a sufficient density of electrons and holes available. In the gas laser case, we looked at how many atoms were excited—how many atoms had electrons in higher energy states. In a solid-state laser, the atoms are linked together throughout the material and we must look instead at the number of electrons and holes per unit volume in the central region—their density. This may seem intuitive if you think about the fact that it is necessary to have a sufficiently large number of excited atoms (electrons and holes) in close proximity to allow laser action to be sustained. There exists a certain value for the density, called the threshold density for laser action, above which the laser material contains sufficient electrons and holes to build up a sustained output of laser light. Ensuring a sufficient density, or reducing losses in order to lower this threshold density, is one of the primary goals of a semiconductor laser design.

A Semiconductor Sandwich

These early lasers were not practical at all. They could function only in response to pulsed currents, and the currents had to be at rather high levels. Semiconductor lasers always operate better in extreme cold. These lasers usually lased only at a temperature of around 77 Kelvin, the temperature of boiling liquid nitrogen. Physicists soon found that they could improve efficiency by changing the configuration of the electrical contact on the top surface of the *p*-type layer. Their solution was to make this contact a stripe of metal, about 10 to 20 micrometers in width. The electrical current passing through the contact was more or less confined laterally as it flowed into the material. Instead of exciting the whole area, it excited atoms only in a stripe centered in the device, which produced a laser beam much narrower in the lateral direction, as illustrated in the figure on the next page. The same level of excitation was achieved with less current overall, as was desired. In this way, pulsed laser action

The early semiconductor lasers were simple *pn* junction diode devices, fractions of a millimeter on a side. Current was supplied through a narrow metal stripe laid down photolithographically along a line perpendicular to the two cleaved crystal facets that served as the laser's mirrors. One disadvantage of this simple design was that electrons and holes injected into the interfacial region tended to drift away slightly, limiting the density of light-producing carriers in that region.

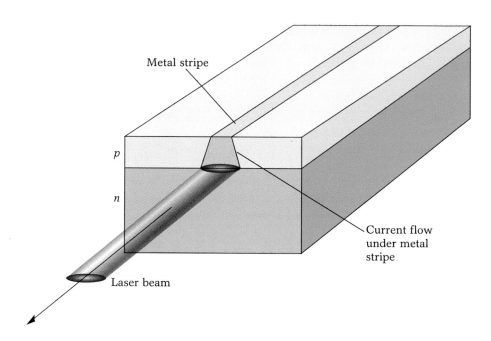

Metal stripe

p

n

Current flow under metal stripe

Laser beam

was obtained at room temperature, although still at very high current levels. Even very short pulses required a current of around 10 amperes. One of the authors of this book, Robert Nahory, made a number of early lasers of this type out of the alloy indium gallium arsenide. These lasers required short pulses of current lasting 100 nanoseconds in the range of 10 amperes up to a whopping 100 amperes at room temperature. The latter were clearly bad devices. For comparison, an ordinary 100-watt light bulb, powered by the standard U.S. 120-volt electric service, requires a current of only 0.8 ampere.

Clearly this type of laser could not be very practical. One problem has to do with the volume occupied by the electrons and holes that generate the photons. When these electrons are forced across the junction interface into the *p*-type material, they penetrate a distance on the order of a micrometer. Thus, as these electrons drop from conduction band to valence band, all the photons are created in a region, 1 micrometer thick, next to the junction interface. The newly created photons stimulate electrons to drop to the valence band and create still more photons. As the photons continually reflect back and forth between the mirror facets, the laser beam is formed in this same region, parallel to the interface.

The large thickness of this region, determined by the electron penetration into the *p*-type material, presents a problem. As small as 1 micrometer seems, a region that thick is simply too large and uncontrolled for the laser to efficiently reach the densities of electrons and holes needed for threshold.

The solution to the problem came along in 1970 with the advent of the so-called double heterostructure (DH) laser, produced independently

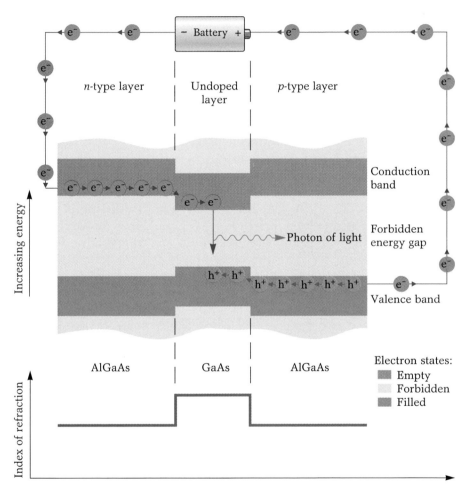

In the double heterostructure laser, first demonstrated in 1970, the electrons entering the central active region from the left and the holes entering from the right are confined by the smaller energy gap (yellow) of the thin undoped active layer material, thus allowing them to reach a greater concentration in this central layer. An added advantage was that the two types of materials in the three-layer structure, here shown as gallium arsenide (GaAs) and aluminum gallium arsenide (AlGaAs), could be chosen such that the index of refraction is higher in the central layer. The higher index of refraction keeps the generated photons guided within this layer, much the way the core of an optical fiber guides light along the center of the fiber.

at that time in two different laboratories in the United States and in the Soviet Union. The first DH design appeared in an article published in the United States in August of that year by four researchers at Bell Laboratories, I. Hayashi, M. B. Panish, P. W. Foy, and S. Sumski. Barely a month later seven researchers at the Ioffe Institute in Leningrad—Zh. I. Alferov, V. M. Andreev, D. Z. Garbuzov, Yu. V. Zhilyaev, E. P. Morozov, E. L. Portnoi, and V. G. Trofim—published a DH design in the Soviet Union. Neither group had known about the other's work. The inspiration behind the DH design, shown schematically in the figure on the preceding page, was to confine electrons and holes, as well as light, in a central active layer by creating a sandwich structure. In the first DH laser, a central active layer of gallium arsenide was situated between two single layers of completely different composition. To do this, the creators of the DH laser relied on their ability to grow precise layers of differing semiconductor materials one upon the other.

Laser designers had already used this capability to grow first a layer of *n*-type material and then, atop it, a layer of *p*-type material. In that case, the semiconductor material was the same for both layers, say gallium arsenide, with only the type of atoms in the minute impurity added to the gallium arsenide changing. Now, however, they needed to grow semiconductor materials of completely different chemical compositions one upon the other. If gallium arsenide was to constitute the active layer, they had to choose a material that would grow with the same crystal structure as gallium arsenide and have approximately the same spacing of atoms. One semiconductor material that meets these two criteria is one in which some of the gallium atoms are replaced by aluminum atoms. This material, referred to as aluminum gallium arsenide or AlGaAs, differs from gallium arsenide in two key respects. First, it has a larger forbidden energy gap. Second, it possesses what physicists call a lower index of refraction. Materials with a higher index of refraction are better able to confine light. Glass, for example, has a higher refractive index than air, and, in the right geometry, it is able to trap light within it. Consider the well-known case of fiber optics. A glass fiber, illuminated by light at one end, transports the light to the other end while permitting virtually none of the light to leak out the sides into the air. The trapped light is obeying a law of physics that states that when light traveling in a material of higher index of refraction approaches an interface with a material of lower index of refraction, it is completely reflected if the grazing angle is less than a certain critical value. If you've ever looked up from under water in a swimming pool, you will have experienced the effects of this law firsthand. You can see out of the pool if you look straight upward, yet if you look out at enough of an angle, peering

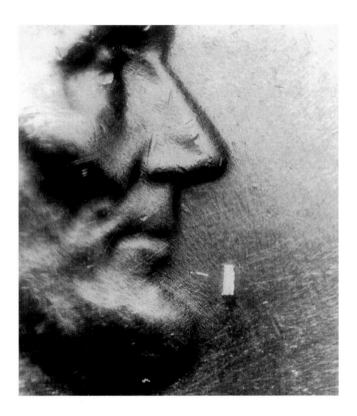

One of the first semiconductor lasers to employ the layered sandwich structure, known as the "double heterostructure," was made by a group of researchers at Bell Laboratories in 1970. Less than a half a millimeter in length, and dwarfed by Abraham Lincoln's profile on the face of a penny, it was the first semiconductor laser to convert electric current into light efficiently enough to run continuously at room temperature.

down the length of the pool, the surface looks like a reflecting mirror. The surface reflects light back to you because the water has a higher refractive index than the air above. Inside a glass fiber, the light beams only approach the fiber's wall at a glancing angle and are therefore completely reflected internally and travel the length of the fiber before escaping.

Just as the glass fiber confines the light, so a slab of higher-index gallium arsenide, sandwiched between two layers of aluminum gallium arsenide, will confine the light within it. To produce the light in the layer, the laser's creators again needed to supply electrons and holes from n- and p-type doped layers respectively. Here they added the two dopants to the two outer aluminum gallium arsenide layers. Now it is clear why the larger energy gap of the aluminum gallium arsenide, mentioned earlier as the first difference between the two materials, is important. Because the gap is smaller in the gallium arsenide active layer, electrons in the n-type aluminum gallium arsenide want to drop down in energy by pouring into the adjacent gallium arsenide. Similarly, the holes from the p-type

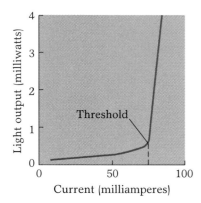

In this example, the light output from a semiconductor laser shows a dramatic jump at about 75 milliamperes, a feature on the curve referred to as the "threshold current." At this threshold, the density of holes and electrons in the central active layer becomes great enough to support the sustained buildup of laser light.

aluminum gallium arsenide prefer to "float up" by entering the gallium arsenide. Again, as in the case of the simple *pn* junction, interface charge barriers will eventually develop to prevent further carriers from entering the gallium arsenide, but with a battery to provide a boost we can keep the electrons and holes flowing indefinitely, producing the glowing slab of light we need for our laser.

This innovative advance in structure made possible continuous laser action at room temperature. The first lasers of this type, created in 1970, operated only for a second or so before the devices burned out, but these results nonetheless opened the way for longer lasting lasers to come. This new structure was one of the major advances in semiconductor laser design and a major step toward the practical application of semiconductor laser devices.

The earliest double heterostructure lasers had active layers with thicknesses of around 1 micrometer, but as these lasers continued to be refined during the 1970s, the active layer thicknesses were for the time being optimized at around 0.2 micrometer. The confinement of electrons and holes in the central gallium arsenide layer increased the density of electrons and holes; as a consequence, the laser reached its threshold of operation with less current. This reasoning suggests that the thinner the active layer could be made, the lower the current required for laser action. Without further sophistication in the design, however, another limitation appears. If the active layer is made too thin, light is not confined but simply leaks away vertically. In this case, the light behaves as though it were attempting to pass through a very narrow slit like the one in the left-hand panel of the figure on page 19, no longer heading in a single direction. Thus the active layer has an optimum thickness, not too thin (to avoid optical leakage effects known as "diffraction") and not too thick (to enhance the density of electrons and holes).

The dramatic operation of a semiconductor laser is illustrated by the graph on this page, which shows the device's light output plotted against the magnitude of electric current that flows through the device. Imagine a scientist turning a knob that increases the current while she measures the light coming from one of the laser facets. At first the light intensity increases slowly until a certain value of current is reached, at the point at which the density of electrons and holes in the active layer becomes large enough to support the sustained buildup of laser light. Above this point the laser turns on, and the light intensity increases rapidly and abruptly. This effect is so dramatic that a scientist carrying out research on such lasers never tires of looking for this threshold for laser action while turning the current knob. The threshold current, along with the steepness of the graph above threshold, is a primary measure of laser quality. The lower the threshold current and the steeper the rise, the better the laser.

Controlling a Laser's Wavelength

Controlling the wavelength of the light emitted by these early lasers was not simple. Typically the laser action took place at a number of wavelengths. A typical spectrum, like that shown in the graph on this page, had one very large peak and several weak peaks. As the current was increased, the auxiliary peaks became stronger until they were comparable in size to the central peak. This sort of spectrum is not desirable for most uses of lasers; only one of the wavelength peaks is generally preferred. The fact that many peaks are observed indicates that multiple waves, or "modes," exist in the active layer of these lasers. These lasers are called multimode lasers. Laser designers needed to change the structural design of their devices so that only one of the modes in the optical cavity was supported, producing single-mode lasers with a spectrum reduced to a single intensity peak at one of the wavelengths.

In the early 1970s optical fibers began to look promising, even though the large losses of light still prohibited practical applications. Hoping that others would find a way to reduce these losses, laser specialists at various research labs began the search for suitable lasers for use with fibers. The existing gallium arsenide lasers performed poorly, but, worse, the wavelength was far from optimum. Optical fiber transmits light with the lowest loss at wavelengths of 1300 nanometers and 1550 nanometers, whereas gallium arsenide lasers operated at about 880 nanometers. New materials had to be created in the laboratory, making use of elements from column IIIB and column VB in the periodic table. The useful atoms are aluminum, gallium, and indium in column IIIB, and phosphorus, arsenic, and antimony in column VB. Lighter elements in these columns at the upper part of the table tend to give lasers that emit at shorter wavelengths, while the more massive atoms lower in the table tend to give lasers that emit at longer wavelengths. By trying out combinations of one or more elements from column IIIB and one or more elements from column VB, researchers learned how to make alloy layers from an alphabet soup of materials useful for new kinds of lasers operating at the right wavelengths for fiber optic communications. These materials are such alloys as aluminum gallium arsenide antimonide (AlGaAsSb) and indium gallium arsenide phosphide (InGaAsP). The latter III-V alloy has emerged as the material of choice for most fiber optic applications.

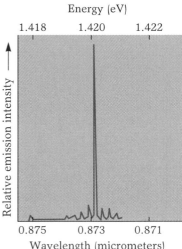

An early semiconductor laser emits most of its light in the narrow band of wavelengths indicated by the central peak. The existence of multiple smaller side peaks, however, indicates that lasing might also occur at alternative wavelengths. A shift from one wavelength to another in a laser's output is highly undesirable for lasers in a fiber optic communications system. Many of the subsequent refinements in semiconductor laser design were aimed at eliminating this shortcoming.

Fiber Optic Communications

Fiber optic communications is the major application of semiconductor laser technology, which will form the basic fabric of the coming information age. Using three primary components — lasers, light detectors commonly referred to as photodetectors, and optical fibers — the technology can harness light for transmitting data, images, and sound at high speed.

All two sites need to communicate is the following basic setup: a laser transmitter at the sending site, an optical receiver at the receiving site, and a fiber link between. Information is sent digitally as a stream of light pulses generated by the laser, passed through the fiber, and read by the photodetector at the far end. But first the data, sound, or images must be encoded into digital form, by means of one of several different coding schemes. In one scheme, each pulse of light is interpreted as a single bit of information, a binary 1. Lack of a pulse, represented by either no light or light at a low background level reaching the photodetector, is interpreted as a binary 0. A stream of timed laser pulses can readily carry a large amount of computer data to faraway places. Digital audio, used in compact discs, and digital video, used in new digital video discs and high-definition television, are readily compatible with fiber optic technology.

The standard telephone, however, still transmits voices in analog format rather than digital. Signals in analog format are made of continuously varying voltages or sound waves. Fortunately, so-called sampling techniques can readily transform such signals into digital format. The concept is to look at the value of the voltage signal periodically at very short intervals, convert that value into a digital number formed of a series of 1s and 0s, and then string all these digital numbers together into a continuous digital stream. Electronic circuits sample the sound of a symphony orchestra, for example, at the rate of about 40,000 times each second to make up a set of digital data for storage on compact discs. A sequence of pits on the disc represents the binary 1s and 0s of the digital signal. A mini-optical communications system within each CD player reads compact discs for playback by shining a semiconductor laser beam onto the spinning disc and sensing pulses of light flashing off rapidly passing pits. A similar sampling technique transforms telephone conversations into bits that can be transported over fiber as pulses of light. The process is reversed in the receiver to recreate analog sound waves.

Today's fiber optic systems send and receive information at rates of 45 million bits per second or faster. Systems now being tested operate at extremely fast speeds, transmitting up to 1 gigabit per second (1 gigabit =

1 billion bits = 10^9 bits). For even higher speeds, laboratories are developing multiwavelength systems, which couple the light from many lasers of different wavelengths into the same fiber. Though transported together, the different color beams remain independent. Laboratory models already exist of chips containing 40 lasers of different wavelengths integrated into a single package. This is a time of innovation and rapid technological advance, notwithstanding the fact that the deployment of basic fiber is still in the early stages.

An interesting exercise provides a feel for these digital speeds. We begin by first digitizing some standard information. Consider the Old Testament of the *Bible:* in a standard version there are 70 lines of text on each page, at 100 characters per line including spaces. That gives $70 \times 100 = 7000$ characters per page. There are a total of 635 pages in the particular version in front of us, giving a total of $635 \times 7000 = 4,445,000$ characters in all. These characters can be typed at a computer keyboard and saved in digital format as a file. In standard coding, each character is represented by 8 bits (ASCII code), so that a total of $8 \times 4,445,000 = 35,560,000$ bits must be stored. Suppose such information is stored electronically as a file somewhere in a network ready for remote access. Using even a basic fiber optic network operating at 45 megabits per second, we could electronically download this file in less than a second!

A similar analysis of a standard 20-volume encyclopedia yields 1 gigabit for storing the text, with another gigabit or so required for storing the images. Using a fast fiber network operating at a gigabit per second rate, we could download the entire encyclopedia in seconds. Of course, we would then want to include additional animated sequences and video clips, requiring somewhat more download time. And we would soon demand even faster systems for downloading full-motion video files.

Some loss of light from a fiber from scattering and absorption in the glass is inevitable, and after traveling some distance an optical signal will become too weak for the photodetector to read. How far can a laser signal travel in a fiber and still be interpreted? The current laboratory record is a distance of 200 kilometers. In practice, distances are shorter. On land, it is not uncommon for uninterrupted lengths of fiber to extend 10 kilometers. The optical undersea cables that now stretch around the world have segment lengths between 50 and 100 kilometers. At points between these lengths there are "repeaters," optoelectronic systems that amplify and recondition the signals for the next fiber length.

A standard repeater, used in fiber systems operating at wavelengths of 1300 nanometers, consists of a photodetector, an electronic amplifier,

Global communications companies employ specialized ships to lay undersea fiber optic communications cables. Here a cable is eased off the ship through a special device in the bow, as part of the laying of a complete cable system across the entire Atlantic Ocean.

Undersea fiber optic communications cables require devices known as "repeaters" to boost the light pulse signals as they grow dim every 50 to 100 kilometers or so. New repeater systems, such as the one shown here, boost the optical pulse signals by passing them through a section of fiber impregnated with erbium atoms pumped to an excited state. These excited erbium atoms release photons when triggered by the incoming dim pulse and increase its intensity. The process is closely akin to the amplification by stimulated emission that is at the heart of all lasers.

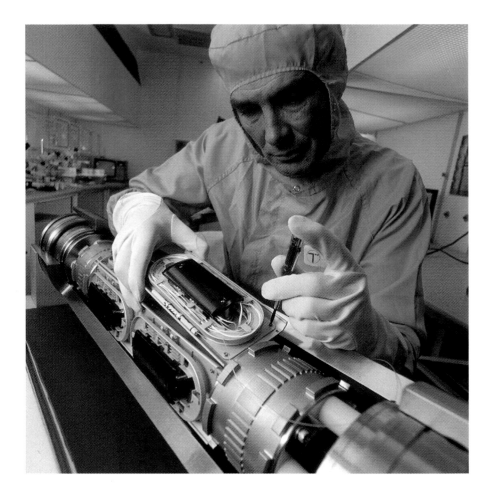

and a laser. The optical signal is received, converted into an electronic signal, amplified, and then sent to the laser for regeneration as an optical signal for transmission through the next length of fiber.

A new kind of repeater has been developed for use in fiber systems operating at 1550 nanometers. This type of amplifier operates directly as an *optical* amplifier. Recall that the stimulated emission process is a way to clone photons. Harnessed appropriately, the process is just what is needed for an optical repeater. Fiber amplifiers consist of a 10-meter length of fiber that is doped with the rare-earth element erbium. The orbitals of the erbium atoms are spaced just right for stimulated emission of photons at 1550-nanometer wavelengths. The erbium atoms of course need to be excited to put electrons into the required orbitals, a step that is accomplished with a semiconductor laser and a fiber that is connected into the main fiber by a "y" junction like the merging on-ramp of a highway.

Placing one of these fiber amplifiers in line with a standard fiber provides the needed amplification. As photons of 1550-nanometer wavelength pass from a standard fiber into the erbium-doped fiber, they begin to "see" excited erbium atoms embedded in the fiber. Through the process of stimulated emission, they cause electrons in excited orbitals of erbium to jump to lower energy orbitals, emitting a photon clone in the process, just as takes place in the neon gas or semiconductor laser. We end up with more photons in each pulse of light, and our optical signal is amplified as desired.

These new amplifiers are now being installed as fiber optic communications systems continue to be deployed on land and under the sea. The promise of these systems is tremendous; they will certainly impact our lives in ways that we have not yet completely fathomed. The development of fiber optic communications has even been called the greatest revolution of the twentieth century. Time will tell.

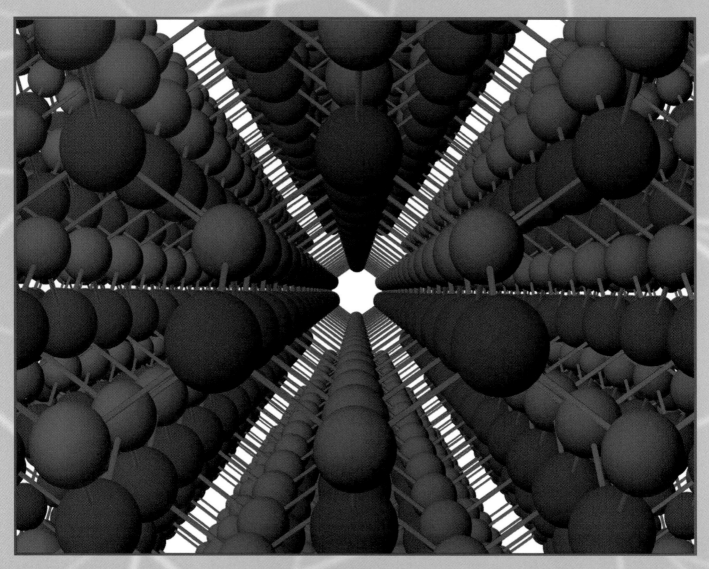

Rows of gallium (purple) and arsenic (green) atoms, together with the regular open channels which separate them are apparent in this close view of a portion of a model of a gallium arsenide crystal. It is along this direction that the transmission electron micrograph on page 23 was taken, and in that image each spot in the gallium arsenide layers corresponds to one of the rows seen above.

6

Growing
Semiconductor Crystals

Even the simplest semiconductor laser is built from a crystal that consists of layers of different semiconductor materials. And now lasers of far more complex design fill a host of specialized niches ranging from units to provide the light that powers optical fiber amplifiers all the way down to surface-emitting lasers small enough for a million to fit on a chip the size of your fingernail. Yet no matter how complex, these lasers all are fabricated from layers of extremely perfect crystals of semiconductors. They could not exist were it not for the existence of a technology able to create nearly perfect crystal layers of precisely controlled thicknesses and chemical compositions. Sophisticated techniques allow scientists to

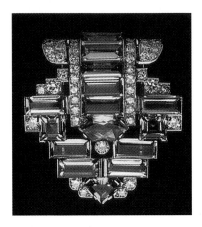

The diamond and aquamarine gems in this classic 1937 Cartier clip brooch are examples of nearly perfect single crystals found in nature. The clearly visible angular facets on each stone are a macroscopic indication of corresponding planes of atoms in the underlying crystal structure at the microscopic scale.

tailor the physical properties of these fabricated crystals in virtually arbitrary ways, producing crystals with abilities inconceivable in the crystals available to us in nature.

What Is a Crystal?

The word "crystal" often evokes the mental image of a ruby or sapphire, sparkling in the intense light of a jeweler's window. And in fact gemstones are crystals, some of the most dramatic crystals found in nature. But so, too, are such mundane objects as a sheet of aluminum foil, a rock in your backyard garden, and even a grain of common table salt. These objects are all solids, but so is the glass in your window pane or the plastic of a milk container, materials that are not crystalline at all. All solids are built up as a rigid arrangement of their constituent atoms, linked by chemical bonds. What distinguishes a crystalline solid from a noncrystalline solid is the particular arrangement of the atoms within the solid.

In noncrystalline solids like glass and plastic, the atoms are arranged, as they are in a liquid, in a rather random way. Imagine the atoms in such a solid as marbles being thrown into a large sack, shaken up, and allowed to collect at the bottom. There is no reason to expect that they will arrange themselves in an orderly way. This sack full of scrambled marbles is in fact a good model for a noncrystalline or "amorphous" solid. One of the first classic models of the random arrangement of atoms in a liquid, in fact, was published in the early 1960s by the British scientist John Bernal of the University of London, who created his physical model in much the way we have just described. He and his colleagues filled a large balloon with three thousand ball bearings, shook them thoroughly, and then poured paint in between them to freeze their positions in place. In a heroic effort Bernal and his coworkers disassembled each sphere out of the sticky mess one at a time after precisely noting its position. In this way they came up with a set of (x, y, z) coordinates describing the structure of a "typical" liquid or, in fact, an amorphous solid. This set of coordinates has been used extensively since then for all kinds of numerical modeling of various properties of an amorphous solid. The ball-and-stick representation of that original model shown on the opposite page, assembled by Bernal himself, provides a nice visualization of what the structure of a noncrystalline material might look like.

The atoms in a crystal, on the other hand, are arranged in precise positions with respect to one another, in a repeating pattern that could in principle continue forever. The angles of the bonds connecting these repeating series of atoms continue in a regular, predictable repeating pattern as well. The figure on page 116 shows a simple model of such a

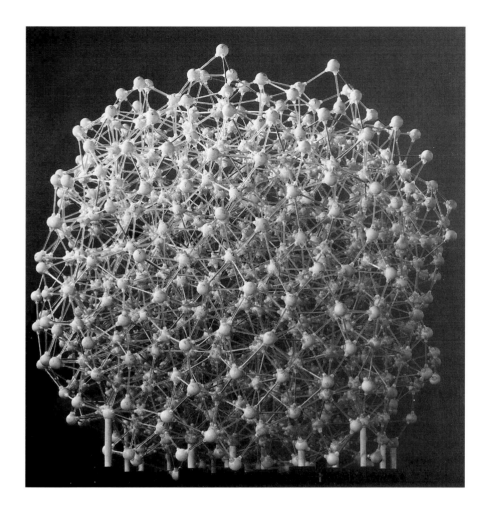

Proposed in the early 1960s, Bernal's model showing a typical configuration of the atoms in a liquid is based on the positions his group painstakingly measured of ball bearings in a random heap compacted inside a balloon. The model has since been used extensively as a representation of the lack of order in a noncrystalline, or "amorphous," solid. Like our earlier models of the structure of silicon and gallium arsenide, on pages 77 and 97, the model represents each atom as a small ball, connected to its nearest neighbors with sticks that represent the bonds holding the atoms together.

crystal, with each atom represented by a solid sphere. The crystal structures of silicon and gallium arsenide seen in the previous two chapters are additional examples of crystals. These semiconductor crystals were a bit more complex than the one shown on page 116, but the repeating nature of the structure is evident in all cases. The fundamental property of crystals is that the positions of all the atoms can be specified completely by designating each atom's position within a subunit of the crystal known as the unit cell (such as the cube shown on page 77) and describing how an unlimited number of these cells can be stacked one upon the other to completely fill the space of the crystal.

Although examples of amorphous solids are widespread in nature, the reader may find it surprising that they are the exception rather than the rule. Most solids take on a crystalline state, a fact that can be

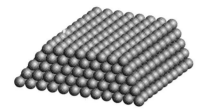

A crystal consists of atoms packed in regular repeating patterns. The simple close-packed structure shown here, for example, represents the particular three-dimensional pattern in which the atoms of many metals are arranged. This "space-filling" model employs an alternative way of depicting atomic positions: it represents each atom as a large sphere whose radius is just half the distance to its closest neighbor. Such a representation is better than a ball-and-stick model at depicting how the atoms fill space, although it is less effective at showing the directions of bonds to nearest neighbors.

understood at least in part from our discussion in Chapter 4 of the bonding arrangements that achieve stable filled electron shells. There is a strong driving force for getting the short-range order—the arrangement of the nearest neighbor atoms to which a given atom is bonded—"correct" in an electron shell-filling sense. And the formation of a crystal, which imposes a longer-range order, is an ideal way to ensure that this requirement for short-range order is met for each atom.

It is possible to form a noncrystalline, three-dimensional network of atoms that meets these short-range order requirements for most of the atoms in the network. But it is difficult if not impossible to meet this requirement for absolutely *every* atom in such a random network. Thus "defects" result that make the noncrystalline solid slightly less preferred energetically than its crystalline counterpart. In general, an amorphous solid forms in nature not because it represents a state of lower energy but because, in the process of forming the solid, the atoms did not have enough time to move around and find their proper lowest-energy crystalline positions. Under most situations and for most solids, the natural solidification process allows more than ample time for the crystalline arrangement to form. Although pure semiconductors such as silicon and gallium arsenide do not occur naturally, when we form them in the laboratory by solidifying the right combinations of atoms, they do form into crystals. And this is significant in that even the very small number of bonding defects that would inevitably form in a noncrystalline arrangement would provide highly undesirable uncontrolled sources and sinks for the extra electrons that we are trying so carefully to control through the use of doping.

Single Crystals

Just because a material has the propensity to adopt the crystalline state as it solidifies does not guarantee that an entire mass of that material will turn into a single crystal. Think of the ice crystal forming in an ice cube tray filled with water, cooling in the freezer. Once the temperature of the water drops below water's freezing point of zero degrees Celsius, small ice crystals begin to form, but they form simultaneously at multiple places within each pocket of water. As the freezer continues to remove heat from the water, more and more of the liquid turns to ice, the result of water molecules joining onto one of the crystals already present. Thus each initial crystal in the cube grows larger, consuming the water around it, until it runs into adjacent crystals and stops growing. The result, then, is an array of small crystals joined together to form the macroscopic object that we refer to as a cube of ice. Such a material is termed "polycrystalline," meaning "many crystals," and, with some notable exceptions such as gems, virtually all naturally occurring crystalline material that is large enough to pick up and handle is in fact polycrystalline. Within each crystalline domain, or "grain," all the atoms are arranged in their mathematically described repeating positions to form a perfect single crystal. But since each of the initial crystallites from which the individual grains grew is oriented at random with respect to the others, the regimented order inside the crystal will be broken as we pass across the boundary separating two adjacent crystalline grains, referred to as the "grain boundary."

The figure on this page shows a model of how the atoms might appear at an idealized grain boundary. The regular arrangement of neighboring atoms, and hence the interatomic bonding, is disrupted. The regular number of complete bonds with neighboring atoms is no longer present for those atoms at the grain boundary.

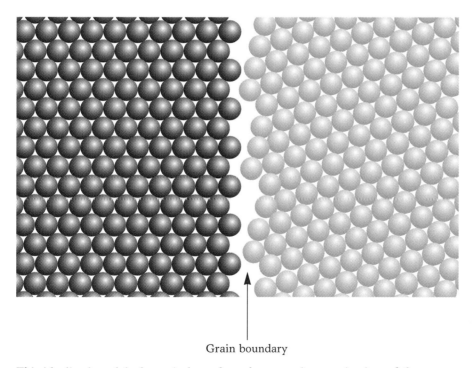

Grain boundary

This idealized model of a grain boundary shows a microscopic view of the atomic arrangement at the boundary between two adjacent crystals, or grains. Although atoms in the two grains are shown in different colors for the sake of clarity, the adjacent grains in an actual polycrystalline material would consist of atoms of the same types. Some partial form of bonding can occur across the interface, but many atoms are left without the desirable number of nearest-neighbor atoms to bond with.

Metallographers have used staining and special imaging techniques to develop the false-color contrast evident in this microscopic view of a piece of brass, designed to highlight the various small miniature crystals, or "grains," in a polycrystalline material. Each grain is a near-perfect crystal on the order of a tenth of a millimeter across, and each is oriented in a random direction with respect to its neighbors.

Like freezing water, molten semiconductors naturally cool into a polycrystalline state. The problem with a polycrystalline semiconductor material is that at the grain boundaries one finds disrupted bonds that do not link cleanly with the material at the other side of the boundary. Often referred to as "dangling bonds," these unlinked bonds have the same ability to add or subtract electrons as intentionally inserted impurities do, but the number of such disrupted bonds and whether they accept or donate electrons is in practice impossible to control in the way that one can precisely control the doping of a material with impurities.

So we must add to the demanding specifications required for a semiconductor device. Not only must the manufacturer exercise incredibly tight control over the potentially small amounts of impurities (sometimes at the part per billion level), but no grain boundaries must be present as well. In practice this is accomplished by constructing our semiconductor laser on top of a wafer that already consists of one single crystal of material. Furthermore, the completed laser needs to be a single crystal consisting of a sandwich of multiple layers, including the active layer and two or more confining layers, that maintains the crystal structure across the boundaries between layers. Although nature does manage from time to time to create single crystals—rubies and sapphires are examples—that are macroscopically large, it hardly ever creates single crystals of layered materials such as we require for our laser. So how have scientists achieved this feat? The materials must be man-made, and they must be created in a series of painstaking steps.

Bulk Crystal Growth

The first step is to prepare a starting template, consisting of a single crystal, onto which we will subsequently lay down layers of different material. This starting template, known as the "substrate," is made entirely of a single material. Its crucial feature is that it is one single crystal across its entirety, with no grain boundaries.

The techniques for fabricating such large single crystals were developed during the years around World War II, when work on early semiconductor devices, most notably the transistor, was beginning to flourish. These early devices were made from large chunks of crystals; refinements in device capabilities went hand in hand with refinements in bulk crystal growth. By the time layered growth techniques, which are essential for the creation of the semiconductor laser, were being explored a decade or two later, the art of creating near-perfect bulk crystals was highly developed. Today virtually all the semiconductor devices in large-scale manufacture, from computer chips to the mass-produced lasers used in audio CD players, are formed as layers on top of bulk crystal templates in the form of thin wafers. The major difference between large-scale production facilities and the scientist's laboratory, where entirely new devices are created, is simply in the scale of the operation. A semiconductor foundry adds layers to hundreds of wafers an hour, whereas a laboratory scientist may add layers to a handful of wafers a day. Both, however, employ the same commercially prepared bulk crystal wafers as the starting template for crystal growth.

The process of creating this single-crystal template is simplest if it is made purely from silicon or some other unmixed element. The bulk crystal grower starts with a lump of pure silicon, which is a polycrystalline material, and cuts from it a piece that is smaller than the size of the average grain, so obtaining a single tiny crystal. This crystal will serve as the "seed" from which he grows a large crystal block of silicon. The seed is inserted into a large pool of melted silicon, cooled almost to the point of freezing, called the "melt." In fact, a temperature gradient set up in the melt ensures that the coolest part, the part closest to the solidification point, is on the surface, right at the top where the seed is to be inserted. The temperature elsewhere in the melt is high enough that the atoms, although in close contact with one another, have enough energy as they slide past one another to break bonds as fast as they form. Thus the liquid atoms avoid forming the steady long-lived bonds that make atoms coalesce into a solid. At the top of the melt, however, the temperature is cool enough to allow permanent bonding to occur. The crystal grower lowers the small single-crystal seed to just touch the surface at this point. Then the melt is cooled a bit more, and the solidifying silicon atoms, finding a point at which they can bond with other silicon atoms, settle onto the surface of the seed, adding incrementally to its preexisting crystalline pattern. In this way, the seed grows progressively larger.

In order to keep the process going, the crystal grower slowly withdraws the seed upward from the melt at a rate just equal to the rate at which the interface between the solid and liquid is growing downward. As he lifts the seed, he slowly rotates it, to ensure better uniformity in the melt. This technique, known as Czochralski crystal growth, is a widely used bulk growth technique, and in addition to being the mainstay in the preparation of semiconductor substrate wafers, it is also the technique often used to create the crystals such as the ruby used as the light source in solid-state ruby lasers.

The outer edges of the melt container must be kept slightly warmer so that new crystals do not start to form at the periphery. In fact, wherever there is contact between the solid and the liquid, some cooler atoms within the liquid are crossing the boundary to join the solid, and some

The first stage in the manufacture of a semiconductor device such as an integrated circuit or semiconductor laser is the creation of one large block of single-crystalline material, referred to as a "boule." From this boule will come the wafers that serve as the substrate template for the deposition of further layers. Here, in a technique known as Czochralski crystal growth, a tiny single crystal of silicon is suspended in a pool of hot molten silicon and slowly withdrawn upward as the crystal grows from the "melt." The resulting boule can be many centimeters across and as much as a meter in length.

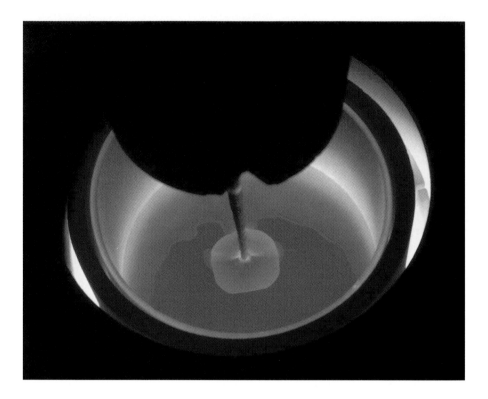

warmer atoms within the solid are using their extra bit of energy to break off the edge of the solid and join the liquid. The rates at which atoms enter and leave the solid exactly balance each other, in what is termed "equilibrium," at the precise temperature that is the freezing point of the material. The art of crystal growing is to reach this point at the exact edge of the rising crystal, so that just enough heat is removed from that crystal to tip the balance in favor of the atoms moving from the liquid to the solid. Maintaining all the proper temperature gradients in time and space, and spinning and withdrawing the growing crystal at rates appropriate to those gradients, is as much an art as it is a science. If the "artist" does everything "just right," a long, round single crystal emerges from the melt, having the overall shape and dimensions of an uncut cylinder of salami.

As a final stage, a technician slices the cylinder perpendicular to its axis into thin disk-shaped pieces using a specialized diamond saw, in a manner analogous to the way a butcher's clerk might slice the salami in a grocery store into a pile of individual slices. Each slice, referred to as a "wafer," is then polished to a flat mirror surface and chemically etched and cleaned to an atomic cleanliness. It is upon this single crystal that the series of layers needed for the laser or other electronic device are deposited.

The sequence of steps is much the same for creating a substrate from a compound semiconductor such as gallium arsenide (GaAs), the first stage in the fabrication of many semiconductor lasers. The key difference is that the melt must contain a mixture of the two elements (gallium and arsenic in this case). The crystal grower needs to carefully control what the chemist refers to as the "stoichiometry," or fractions of

A bulk single-crystal boule of gallium arsenide (right rear), grown by the Czochralski method, is machined into smooth cylinders (left rear) and then sliced like salami into thin disk-shaped wafers here 50–100 millimeters across and half a millimeter thick. Then, the two flat surfaces of each wafer are mechanically and chemically polished to a mirror finish (front). It is upon such surfaces that the subsequent growth occurs of the layers which comprise the device such as a semiconductor laser.

the different types of atoms in the compound. When either of the two atoms is present in extra amounts, those atoms will incorrectly occupy sites normally filled by the other type of atom, or even force themselves mistakenly into small open spaces between correctly positioned atoms. When either type of atom is present in less than the required amount, there will be "vacancies" on the crystal lattice where an atom was supposed to reside. In any of these cases, the number of electrons contributed to the crystal differs from the proper number required to precisely fill the lower valence band without unintentionally adding any excess electrons to the upper conduction band. In effect, these imperfections act as unintentional dopants, and to the extent that their numbers are difficult to control, they are considered undesirable.

A 50:50 mixture in the grown solid is achieved simply by starting with equal amounts of gallium and arsenic atoms in the melt. The relative amounts of dopant atoms, however, needs to be finely adjusted to take into account their differing propensities for bonding onto the growing crystal. Those differing propensities are also sensitive to conditions such as temperature and pressure, so calculating the initial proportions becomes in principle a rather complex problem and often the empirical approach of trial and error is used to accomplish the last bit of fine tuning. Add to this the fact that arsenic evaporates very quickly at the growth temperature of gallium arsenide, and you realize that the entire procedure must be carried out in an enclosed, carefully pressurized vessel to prevent the evaporation of arsenic from upsetting the fine balance in the amounts of gallium and arsenic atoms. Such problems are still more aggravated for the growth of indium phosphide (InP), since phosphorus evaporates even more quickly than arsenic. It is no wonder that bulk crystal growth is considered as much an art as it is a science.

Epitaxy

So we have seen thus far how a reasonably perfect single crystal of starting material such as silicon, GaAs, or InP can be grown, sliced, and polished to produce wafers of these single-crystalline materials. To make even the simplest semiconductor laser, the manufacturer must grow a series of thin layers on top of one of these polished wafer substrates. The trick is to use the arrangement of the perfectly positioned atoms in the substrate as a pattern for the further growth of the superimposed layers. This process of extending the crystal structure of the underlying substrate material into the grown layer is called "epitaxy." The new, growing crystal takes both the positions of its atoms as well as their bond angle arrangements from the crystal beneath. The version of the technique called homoepitaxy is used to grow a layer of exactly the same crystalline material onto the substrate, as when one grows silicon layers doped with slightly different impurities upon a silicon wafer (for example, growing an *n*-type layer on a *p*-type substrate). In this case, the layers, despite subtle differences in impurity content, are composed of the same basic material.

Heteroepitaxy, on the other hand, is the technique used to grow a layer of different material on the substrate. The necessary condition for the layer growth to be deemed epitaxial is that the overgrown layer take its crystal orientation, its "clue," from the substrate. Hence, in principle, two crystals with, say, an overall cubic arrangement of atoms could be grown heteroepitaxially, one upon the other, even though their atomic positions differ within the cubes, as long as the cubes of the overgrown material were oriented by the cubes of the substrate. In practice, controlling the growth of materials having substantially dissimilar crystal structures is difficult, and although a small research community is grappling with this problem, the production of the complex semiconductor structures required for even the most sophisticated semiconductor lasers relies almost exclusively on the heteroepitaxial growth of materials that, though they may be composed of different atoms, possess identical crystal structures, and by and large even the same distances between atoms within the structure, a characteristic known as their "atomic spacing."

This requirement that two crystals have the same structure and atomic spacing may seem extremely limiting, and in a broad sense it is. But it is not uncommon for an entire class of semiconductors to possess the same crystal structure. For example, virtually all the III-V compounds used extensively in the construction of semiconductor lasers grow in the structure we have seen earlier on page 97. All that differs among these compounds is the atomic spacing, and in some cases, referred to by crystal growers as "lattice matches," even that spacing is nearly identical.

The most widely exploited lattice match is that of gallium arsenide, GaAs, and aluminum arsenide, AlAs. Both compounds have the same structure, and their atomic spacings, or "lattice constants," match to within about a tenth of a percent. Hence one can readily grow AlAs on GaAs heteroepitaxially. AlAs can be thought of as GaAs, as depicted on page 97, in which we have replaced each of the gallium atoms with aluminum atoms. But one can instead do a partial replacement, replacing, say, half the gallium atoms with aluminum atoms. The result is an aluminum gallium arsenide alloy (AlGaAs), sometimes denoted as $Al_{0.5}Ga_{0.5}As$, where the subscripts on the Al and Ga indicate the fraction of the combined Al and Ga sites occupied by the respective atoms, a 50:50 mixture in this case. Materials scientists refer to such a mixture of three elements as a "ternary" alloy. Since both GaAs and AlAs are lattice

Alloys and Lattice Matching

A diagram such as the one in this box may help the reader visualize how laser designers create layers of differing energy gap yet similar lattice constants. This is a plot of energy gap versus lattice parameter (the spacing between atoms, or what is sometimes referred to as the material's "lattice constant") for four semiconductors of the III-V type, namely gallium phosphide (GaP), gallium arsenide (GaAs), indium phosphide (InP), and indium arsenide (InAs). Together these simple two-element compounds offer the laser designer only four possible energy gaps, and they cannot be lattice matched to form a layered heterostructure, as each has a different lattice parameter. Each compound is represented on the diagram by a single dot. For example, GaAs has an energy gap at 1.42 eV and a lattice parameter of 5.65325 angstroms.

Ternary alloys, composed of three elements, offer more choice of energy gap, but only rarely does one of these alloys lattice match to a binary compound. The lines connecting the dots in the

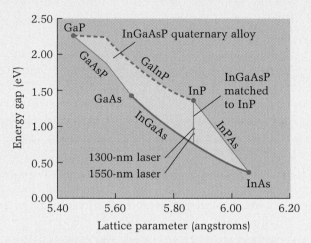

A diagram such as the one shown above relates the energy gap and lattice constant of a class of III-V semiconductors and provides a useful tool to laser designers. Using the diagram, they can determine the composition of a given layer such that its lattice parameter matches the chosen substrate and its energy gap is appropriate to that particular layer's function in the overall heterostructure laser design.

matched, meaning the gallium and aluminum atoms in this crystal structure are the same size, so are all the $Al_xGa_{1-x}As$ alloys in between. And since AlAs and GaAs differ in crucial semiconductor properties such as their energy gap, we can lay down layers having differing amounts of aluminum substituted for gallium and obtain the kind of layered energy gap structure that is essential for the fabrication of a semiconductor laser.

For two compounds such as GaAs and AlAs to have the same lattice constant is a fortuitous advantage, but a rare one. The existence of such a match is unlikely among the other so-called "binary" compounds such as InP, InAs, and GaP that are available by mixing any single Group III element with any Group V element. If we start with an InP substrate, and we want to grow a layer with a different energy gap, it turns out that any element we alloy with either the indium or the phosphorus changes the compound's lattice constant. We can't grow either GaAs or diagram correspond to the relevant ternary alloys. For example, the curved line from GaAs to InAs plots the properties of various compositions of the InGaAs alloy. As the composition is changed, both the energy gap and the spacing of the atoms changes. Thus moving from the GaAs point toward the InAs point, the energy gap gets smaller while the atoms become farther apart. Similar relationships exist for the three other ternary alloys GaAsP, GaInP, and InPAs.

Truly flexible lattice matching becomes possible with quaternary alloys. The entire area inside the lines represents the quaternary alloy InGaAsP. By adjusting composition, crystal growers can prepare materials that fall at any point in the area, giving a wide range of possibilities for energy gap and lattice parameter. One possibility is to prepare quaternary alloys with different energy gaps but the same lattice parameter, such as the alloys that lie along the vertical line extending down from the InP dot in the diagram. This particular set of quaternaries has a lattice parameter identical to that of InP, but the energy gaps range from 1.35 eV at the top of the line to 0.75 eV at the bottom. This uniformity of lattice parameter is precisely what we want for growth of double heterostructure lasers on InP substrates. We can choose InP as both substrate and confinement layer material, then we can choose any InGaAsP composition on the line for our active layer. For example, the two dots in the diagram locate the compositions used for InGaAsP active layers that emit infrared light at the wavelengths 1300 nanometers and 1550 nanometers, the standards in fiber optic communications.

This type of diagram is useful as a tool for designing new structures. For instance, you can easily see that InGaAsP layers could be prepared for lattice-matched heterostructures on GaAs substrates, but not on InAs or GaP substrates. Bulk substrates of the ternary alloy InGaAs have recently become available, which you can see opens up new regions for matched epitaxial structures.

InAs on InP, for GaAs has a lattice constant 4 percent smaller than that of InP and InAs has one 4 percent larger. What we can do, however, is choose an alloy that mixes the two, $In_{0.5}Ga_{0.5}As$. This alloy composition precisely lattice matches the InP, yet has a different bandgap.

Although this clever trick solves the problem of lattice mismatch, it does so for only one particular material, $In_{0.5}Ga_{0.5}As$. But then we would only be able to make a laser with a wavelength of 1600 nanometers, characteristic of $In_{0.5}Ga_{0.5}As$. What if we wanted lasers of other wavelengths? The solution is to add yet another element. Phosphorus, for example, tends to raise the energy gap, but it lowers the lattice constant because it is a smaller element. So the crystal grower adds phosphorus to the InGaAs to bring the bandgap up and simultaneously counters the side effect of a shrinking lattice constant by replacing more of the smaller gallium atoms with larger indium atoms. The result is an alloy with four components, a "quaternary," $In_xGa_{1-x}As_yP_{1-y}$. By appropriately controlling the fractional mixture of the Group III atoms and the Group V atoms, it is possible to form alloys with both a range of lattice constants and a range of energy gaps. In particular, it is possible to form a continuous series of alloys with varying energy gaps that all match to a given lattice constant. So with a little cleverness and a careful juggling of the amounts of four elements simultaneously, the desired flexibility is achieved.

This flexibility in choosing the energy gap arbitrarily across an entire range allows us to construct truly complex structures, which form the basis for all kinds of advanced semiconductor lasers. Through careful calculation and attention to myriad details, the grower is able to build a single crystal, already difficult to find in nature as attested to by the cost of precious gems, but with a particular property, the energy gap, that can be changed arbitrarily in the growth direction of the crystal, a kind of "dial-a-property." This capability to create crystals unheard of in nature is a rather awe-inspiring testament to how finely tuned the science and art of crystal growth has become. It is not unusual on a typical day for a colleague of mine at Bellcore working on the design for a new semiconductor laser to come to me with a sketch of a crystal he or she would like me to grow. And the drawing may include dozens or even hundreds of layers, each with a specific energy gap associated with it. Within a day or two I can fabricate the crystal, which is most certainly the only such object in the entire world, in what has become almost a routine process. The process has become so commonplace in modern crystal growth that the sublime has become routine.

Epitaxial Growth Techniques

For the next several pages, we'll focus on the techniques used by these modern-day alchemists, the crystal growers, to grow complex epitaxial structures on top of single-crystalline substrate wafers. The epitaxial growth process is in many ways analogous to the formation of bulk crystals from the melt, or even to the growth of the small crystallites of ice in an ice cube tray. In all cases, atoms that have lost energy by cooling find an open site at the edge of the advancing crystal and incorporate into the solid. There is a distinct quantitative difference between the cases, however, in that typical growth rates of bulk-grown crystals are centimeters per hour (allowing the synthesis in a working day of a moderate-sized "salami-shaped" bulk crystal, or "boule," from which hundreds of wafers are subsequently cut), whereas the growth rates of epitaxial layers are a thousand to ten thousand times slower. These slower growth rates are necessary to allow crystal growers to precisely control the thicknesses of layers that may be only a few atoms thick.

There is as well a fundamental qualitative difference between bulk and heteroepitaxial growth. In the case of heteroepitaxial growth, the crystal grower needs to be able to interrupt one flow of arriving atoms and substitute another in order to start forming the next layer. Over the past thirty or so years a number of techniques have been developed to allow the crystal grower to change the flow of atoms to the surface. Each was intended to further refine the control of layered semiconductor structures, although many have since been extended for use in the growth of epitaxial crystals destined for a wide array of applications, including magnetic storage, high-performance optical reflectors, and high-temperature superconductors.

The first heteroepitaxial growth technique developed for the growth of semiconductor lasers was liquid phase epitaxy, or LPE for short. It played an essential part in the development of the pioneering laser structures of the late 1960s and early 1970s. It is also the epitaxial growth technique most similar to the technique of bulk crystal growth from a melt used for producing substrates. In fact, LPE can be thought of as a natural extension of the melt process.

We mentioned above that a crucial attribute of a heteroepitaxial growth technique is the ability to change the composition of the flow (or in more technical jargon, the "flux") of atoms arriving at the growing interface. For the case of a crystal solidifying from the melt, the crystal grower must be able to finely control what atoms are in the molten material immediately adjacent to the growing edge of the crystal. Rapid changes in this flux are impractical in the bulk case where the melt consists of many kilograms of material—hardly something easy to abruptly change. So the LPE crystal grower creates a series of small pools of molten material, each trapped in its own compartment within an apparatus made out of graphite and referred to as a "boat," depicted in the figure on page 128. Part of the boat is a graphite piece called a "slider," shown in gray at the bottom of the apparatus; the slider traps and supports a piece of substrate material, about 1 to 5 centimeters on a side, so that it may be rapidly slid in a horizontal motion from one melt to the next in about a second or less. Surrounding the entire assembly, not shown in the figure, is a furnace that maintains precise control of the overall temperature, for, as in the case of bulk crystal growth, the growth takes place by carefully cooling the melt.

The growth process is quite straightforward. The substrate, which initially is not in contact with any melt (panel 1), is slid into contact with the first melt (panel 2) for a period just long enough to allow the first layer of material to form. The layer's composition is determined by the composition of this first melt. As soon as the layer grows to its proper thickness, the slider is moved to bring the substrate into contact with the next melt (panel 3), and the process is repeated to form the second layer. This sequence continues until all the desired layers have been deposited,

During the process of liquid phase epitaxy, a movable slider on the bottom of the apparatus slides the semiconductor substrate from one "melt," or pool of molten semiconductor material, to the next in a second or less. During the time the substrate spends in contact with each melt, a layer is deposited epitaxially onto the substrate, its composition determined by the composition of that melt, and its single-crystalline alignment from the crystal structure of the underlying substrate. The entire assembly, referred to as the "boat," is surrounded by a furnace whose temperature is carefully regulated in order to properly control the rate at which the melt solidifies.

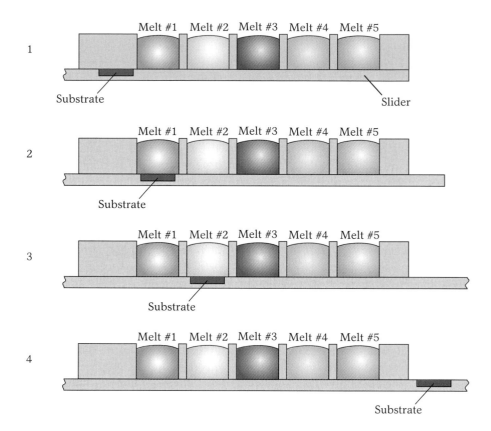

at which point the completely formed crystal is slid away from the melts (panel 4). A new melt is required for each layer having a new composition and/or a change in doping.

With these new techniques for making materials stacked in layers, all kinds of innovative structures could be devised for lasers. Thus engineers were limited not by the material possibilities, but by how far they could push their imaginations. With ample imagination and abundant energy, spurred by a shared excitement, they made rapid progress.

New Techniques, New Designs

It wasn't long before layered structures were no longer simple stacks of planar layers—some odd shapes began to make an appearance, as laser designers found that layers in these shapes could solve nagging problems, earliest among them the reduction of the amount of electric current needed. A metal stripe, even a narrow one, on a standard double heterostructure produced a laser that was still not a very practical device. These lasers required a lot of electric current to compensate for the fact that current from the contact stripe spread out as it flowed into the material, reducing the desired lateral confinement and the efficiency of the laser. But with advances in material preparation, laser designers were able to restrict the lateral current flow so that the current entered right into the

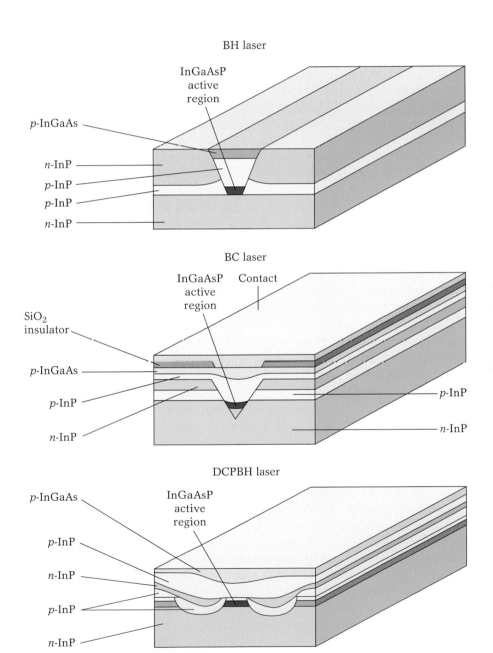

These three semiconductor laser designs of increasing complexity are made possible when an epitaxial growth technique such as LPE is combined with patterned etching techniques. In these designs, some of the layers are grown on nonplanar substrates such as a single central stripe-shaped mesa for the buried heterostructure (BH) laser, a patterned V-groove for the buried crescent (BC) laser, and a pair of broad U-shaped grooves in the case of the double channel planar buried heterostructure (DCPBH) laser. Such lateral patterning creates structures that allow current only into the two-micrometer-wide active layer stripe while blocking its passage around the stripe's sides. As a consequence, these lasers run efficiently at significantly reduced currents.

active layer along the desired stripe connecting the two cleaved mirror facets. In effect, they created a channel for the electricity through the top layers. To create such a channel, the makers of these devices had to combine LPE with another technique, etching. Before this time, all the layers in a crystal were grown in a single process. To prepare the new structures shown in the figure on the previous page required at least three independent steps. The overall idea was to grow a set of layers, etch part of the layers away in some regions, and then follow with another growth step.

The so-called buried heterostructure laser, or BH laser, in the figure on the previous page, was one of the first produced using the new methods. The figure on page 131 shows the steps needed for its formation. In the first step, a complete double heterostructure is grown on a standard wafer. In the second, the etching process is brought into play. A glassy layer of silicon nitride and a layer of organic material called photoresist are deposited on the top surface. Exposure to light tends to break down the molecular structure of photoresist, so that areas exposed to light can be readily dissolved away when dipped into an organic solvent, while the solvent has no effect on photoresist not exposed to light. This property of photoresist is used to control what regions of the layers are etched away. A mask is used for the lateral patterning process consisting of a transparent material, such as glass, on which an opaque material, such as metal, has been deposited in a particular shape. For our BH laser, the mask is a set of metal stripes. Light is shined through this mask which has been placed on top of the photoresist layer, thus exposing it everywhere except under the metal stripes, and subsequently an organic solvent dissolves away the photoresist everywhere except where the metal stripes were.

A special etching process then strips away the silicon nitride, but only where it is not hidden under the remaining photoresist stripes. The rest of the photoresist is then removed, leaving the silicon nitride stripes as the mask. A second etching process removes the semiconductor layers, where not protected by the silicon nitride, down through the active layer, leaving the double heterostructure layers in narrow stripes only about 2 micrometers wide. The entire ridged surface is thoroughly cleaned and then reintroduced into the apparatus for growth of new p-type and n-type InP layers. Note that the silicon nitride layers still on top of the double heterostructure stripes prevent growth on those areas. The silicon nitride is then etched away, and the top surface is metalized for electrical contact. Finally, the wafer is cleaned and sawed into thousands of individual laser devices, each with a buried double heterostructure stripe inside.

Electric current enters the BH laser structure through a broad metal contact on the top, but the current can flow down into the material only at the location of the stripe that was protected in the etching step. The

1. FIRST GROWTH

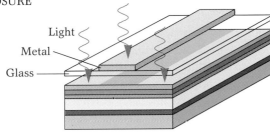

p-InGaAsP
p-InP
InGaAsP
n-InP

2. DEPOSITION OF SILICON NITRIDE AND PHOTORESIST

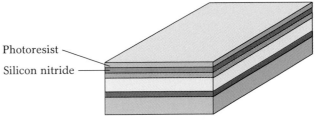

Photoresist
Silicon nitride

3. EXPOSURE

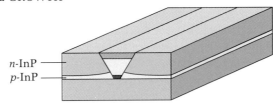

Light
Metal
Glass

4. AFTER ETCHING

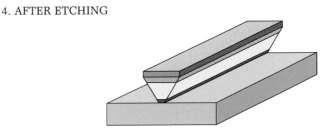

5. FINAL GROWTH

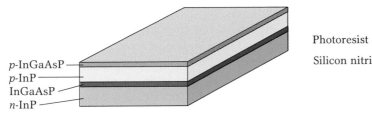

n-InP
p-InP

The processing steps to prepare a buried heterostructure (BH) laser.
1. A complete double heterostructure laser is first grown. 2. A glassy layer of silicon nitride is deposited, followed by a layer of photoresist material. 3. A glass plate with metal stripes, the mask, is placed on the structure and then exposed to light as shown. The exposed photoresist, and the silicon nitride under it, can then be removed leaving stripes. 4. Chemical etching removes the epitaxial layers down through the active region, except for the protected strip mesas, cutting under the silicon nitride mask as shown. 5. Epitaxial layers of p-type and n-type InP are grown, except on the strip mesas where the glassy silicon nitride prevents growth. Finally, the silicon nitride stripes are removed, leaving the planar BH laser ready to be metalized top and bottom and to be mounted on its base for operation.

regrown regions beside the stripes contain a reverse np junction structure that prevents current flow. Thus current flows only in a narrow 2-micrometer width right through the active layer. A typical active layer in the buried stripe region of the device is thus only 0.2 micrometer thick and 2 micrometers wide. This is the shape of the output laser beam as it emerges from the mirror facet, a size that is compatible with the 7-micrometer diameter of an optical fiber core. Low current of only 10 to 20 milliamperes is enough to bring such a laser to threshold.

The BH laser structure was but the first of many designs. Another type of laser designed to confine the beam to a small active layer, also illustrated in the figure on page 129, is the buried crescent, or BC, laser. Note the smiley crescent shape of the active layer. This device also requires two growth steps. The first step creates a *p* layer followed by an *n* layer across the entire substrate. In this case photoresist is next applied to the surface and then exposed under a striped mask and developed to form a pattern of unprotected stripes, the opposite of the case for the BH laser. The masked structure is then etched along these narrow stripes, forming v-grooves whose slanted side walls are crystal facets. After removal of the protective mask material, the surface is cleaned and reintroduced into the crystal growth apparatus for growth of the usual double heterostructure. The active layer automatically forms the crescent shape in the grooves, because of the way that atoms deposit on the crystal facets. The BC lasers produce light only in the crescent-shaped active regions, again allowing the laser to operate with currents in the 10 to 20 milliampere range. The size of the light beam, emitted from a region again only 2 micrometers wide and about 0.2 micrometer in height, is again compatible with the 7-micrometer core of an optical fiber.

One of our favorite names belongs to the so-called DCPBH laser, or double channel planar buried heterostructure laser, illustrated schematically on page 129. To fabricate this structure, a crystal grower first grows a standard double heterostructure on a substrate. A layer of photoresist is deposited and masked so that pairs of grooves, or channels, can be etched closely spaced side by side this time with an etchant chosen to give a U-shaped groove profile. Another growth step is then carried out to provide more layers in such a way that only the region between the two etched channels is active for laser action. The regions to the side have *pnpn* layer structures that block electric current, so that current flows from the top surface down through only the central region in between the double channels. Laser action takes place efficiently, with current thresholds in the 10 to 20 milliampere range, in an active layer once again of very small cross section: 2 micrometers wide by 0.2 micrometer high.

These three types of semiconductor lasers—BH, BC, and DCPBH— are the types now in use in fiber optic communications systems. The best all operate at electric currents around 10 milliamperes and have only a single wavelength peak in their spectra. You can see that they are much improved over the early laboratory lasers that operated with short pulses only at 10 to 100 amperes. These new lasers have very long lifetimes, operating continuously for years.

So many laser structures were tried out during this period of research that many engineers found it difficult to keep up with advances. For a time a common statement heard in the halls outside seminar rooms went something like this: "Today's talk was yet another presentation about a new laser structure with a new name stranger than those I heard about before. Nothing special." It required an expert paying close attention to notice the differences in performance among many of the structures being investigated. An international language of names grew up, recognizable to laser experts around the world. To others the structures all began to look alike.

You have probably noticed great emphasis on materials science and technology, both for the growth of crystalline layers and the fabrication of semiconductor lasers. To make a semiconductor laser indeed requires a variety of skills, in fields as diverse as materials science, clean-room fabrication technologies, and device physics fundamentals. The complex processes necessary to develop new laser devices are carried out by teams of engineers and scientists with skills in physics, chemistry, materials science, and electrical engineering.

Innovations continued as research advanced. Consider some of the enhancements to the DCPBH laser. One new laser device is called a single-frequency laser because it produces a much narrower wavelength peak than earlier devices. In this case the starting substrate is first treated with a mask-and-etch procedure to produce a grating pattern over the whole surface, as illustrated in the figure on this page. A special

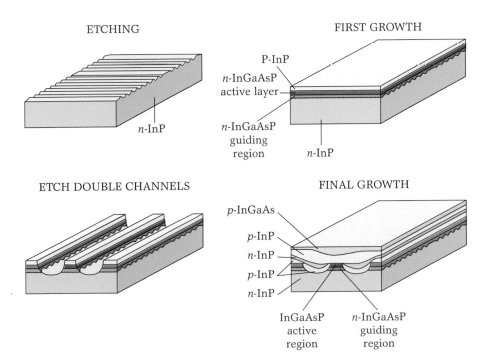

ETCHING

FIRST GROWTH

P-InP

n-InGaAsP
active layer

n-InGaAsP
guiding
region

n-InP

n-InP

ETCH DOUBLE CHANNELS

FINAL GROWTH

p-InGaAs

p-InP

n-InP

p-InP

n-InP

InGaAsP
active
region

n-InGaAsP
guiding
region

A double channel planar buried heterostructure (DCPBH) laser can be improved by depositing it onto a substrate that has been first etched to produce a very fine corrugation, or "grating," in the direction perpendicular to the lasing direction, shown in the upper left portion of the figure. This grating is carefully preserved throughout the subsequent steps required for the improved DCPBH fabrication, illustrated in the balance of the figure, and serves in the final device to lock the operation of the laser at one precise optical wavelength within the laser cavity, determined precisely by the dimensions of the grooves in this initial grating.

resist material is patterned by an electron beam rather than visible light, since the wavelength of light is too large to form such fine features. A very fine set of parallel grooves is subsequently etched only about 100 to 200 nanometers apart with a groove depth of only about 100 nanometers. The trick is to grow the layers of the usual double heterostructure over this grating without erasing it in the process, to produce the structure labeled "first growth" in the figure. The structure next goes to the processing lab for etching of the usual double channels needed for the DCPBH structure. Finally, more layers are grown on top of the etched surface to produce the final single-frequency DCPBH laser.

This laser operates differently from more conventional lasers. Light produced in the active layer travels not only in that layer but also in the corrugated light-guiding layer provided directly underneath, next to the grating. As the light travels, a small amount at a precise wavelength is reflected backward from each groove, all along the grating. Because the grating "selects" a particular wavelength based on the grating's spacing, and light of this wavelength becomes grist for the stimulated emission process, this type of laser generates light of an extremely narrow frequency width. Although the resulting laser beam is strictly speaking not at a single frequency, its range of wavelengths is much narrower than in beams from conventional laser structures that do not have the grating. A single-frequency laser has a frequency width less than 1 megahertz, which is to be compared to the frequency of the light itself at a magnitude of 10^{14} hertz (100 million megahertz).

One further enhancement to the basic DCPBH laser structure is of interest here, the tunable DCPBH laser. All the semiconductor lasers discussed up to this point emit beams of wavelength determined by the material used in the active layers. To obtain a wavelength of 1300 nanometers, for example, the laser designer might choose to create an active layer composed of indium gallium arsenide phosphide with an energy gap corresponding to this wavelength. And that would be that, since the wavelength is decided by this particular combination of atoms.

With appropriate design, however, a given combination of atoms can support laser action over several choices of wavelength within a certain range. A design for such a tunable laser is illustrated in the figure on the opposite page. In the left-hand section of this DCPBH structure, light is generated and amplified as in the conventional laser. To the right there are two new sections integrated together with the DCPBH, formed by the usual etching and regrowth methods. Each of the three sections has its own external wire attached so that current may be applied to each independently. The section on the far right has built into it a grating such as we saw before in the single-frequency laser, with a nearby waveguide layer. It does not have an active layer for generating light. The center

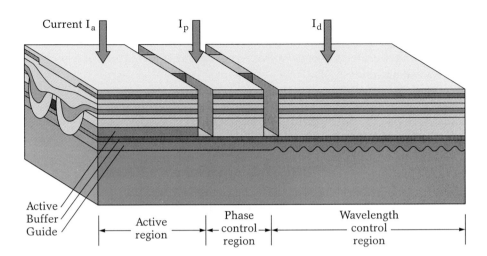

Current I_a I_p I_d

Active
Buffer
Guide

Active
region

Phase
control
region

Wavelength
control
region

This tunable laser, based on the DCPBH structure, is fabricated to have three distinct sections, each receiving a separately controllable current. The left section controls the total light output from the laser, the middle the phase of the internal light field, and the right the exact wavelength of the light emitted. Experimental versions exist that can work at as many as 40 different wavelengths, or channels.

section is just a waveguide section, also with no active layer. A description of this laser in operation demonstrates how a laser engineer with imagination can create remarkable devices.

First of all imagine that you are the controller of this tunable laser. You have a control panel with two knobs on it. The left knob is used to adjust the current flowing to the left section of the laser. The right knob is used to control the current flowing to the right section. You will use the left knob to turn the laser on and off, and you will use the right knob to adjust the wavelength. The laser's center section automatically controls the phase of the light.

By turning the left knob, you first adjust the current to a steady value just above the laser threshold, so that the laser is on but emits a beam of very low intensity. We will call this the off state. Now to send signals out to an optical fiber, you rapidly turn the left knob up and down to create a stream of laser pulses that correspond to a digital code sequence containing intelligent information. You can do this extremely rapidly to generate laser pulses that are only 1 nanosecond in width. You will be sending out data at a rate of 1 gigabit per second (1 billion bits per second). Compare that to your 28.8-kilobit per second modem, which runs more than thirty thousand times slower!

So far this laser is operating at a wavelength determined by the grating in the right-hand section. Its mechanism of action is as follows: Light generated in the active layer of the left-hand section transfers into the waveguide layer and travels into the two sections on the right. In the right-hand section the grating reflects back a precise wavelength determined by the grating spacing. That light reflects back into the active layer of the left-hand section, where it is reinforced by laser action. The process continues so that plenty of light is built up to generate the laser beam, which emerges from the cleaved facet of the left-hand section. You have generated a set of pulses sending out information at a particular wavelength.

Now consider the knob in your right hand. At some point you can change the wavelength by adjusting the current flowing into the right-hand section of the laser. If you increase the current, you will cause more electrons to flow into the waveguide region of the right-hand laser section. The density of electrons in the waveguide will increase, causing the index of refraction for light to change in this layer. As the electron density increases, the index of refraction gets lower.

The index of refraction of a material can be defined as the ratio of the speed of light in vacuum, or air, to the speed of light in the material. The index of refraction is 1 in vacuum and very close to 1 in air. In semiconductor materials the value usually lies close to about 3. When light passes from air into a semiconductor material, the speed of light decreases and the light travels at one-third of its speed in air. There is an accompanying change in wavelength also—the wavecrests bunch up as they travel more slowly in the higher index material, and hence the wavelength of light in a semiconductor is one-third the value in air. As the index of refraction in the right-hand section of the tunable laser changes, the laser light responds by changing its wavelength so that its wavecrest spacing remains in synch with the underlying grating.

So by choosing a given value of current in the right-hand section of the laser, you get a particular wavelength. In practice, engineers have tuned a single laser to generate any of 40 different wavelengths in the vicinity of 1550 nanometers. Each laser wavelength can be switched on and off a billion times a second. They can switch from one wavelength to another just as quickly by rapidly changing the current in the right-hand section. The result is a laser source for a fast multiwavelength fiber optic communications system, able to send secure signals to 40 different receivers, each receiving one of the wavelengths.

This sophisticated example illustrates how a combination of physics, chemistry, materials science, and electrical engineering allows a very sophisticated operation to occur all inside a tiny laser. This laser, like all the others in this section, is smaller than a single grain of salt.

In no way did anyone imagine how much could be done at the beginning of this research. In fact, the evolution of the semiconductor laser is an excellent example of the power of the scientific method. By asking basic questions, and working on the unknown, scientists and engineers were led to the sophisticated laser devices just illustrated. The process has not ended even now. These lasers are a major part of the optical communications revolution of the twentieth century.

Some Shortcomings

Liquid phase epitaxy, when combined with etching techniques, thus opens the door to the creation of these and a wealth of other layered crystals that are unachievable by any bulk growth technique. Yet the technique has its limitations. It is more complicated than you might expect to obtain the proper ratio of elements in each layer. Each type of atom has a different "willingness" to leave the melt and join the solid. Thus the fraction of each atomic component in the deposited layer does not match its fraction in the melt. With care, these differences can be compensated for in the preparation of each new melt, although the process is tricky. Again, we are entering the area of the "art" of crystal growth.

A second complication is that some combinations of atomic elements will not mix when melted together. Instead, like oil and water, they have a natural tendency to separate. The crystal grower is precluded from using these combinations as melts for individual layers, and hence there are certain compositions of layers that, although desirable for a given laser structure, are unachievable by LPE growth.

A third complication has to do with the ability to form layers of extreme thinness. The narrowest layer in the simple double heterostructure laser on page 103 is the central gallium arsenide region, which might be one or two tenths of a micrometer thick. This corresponds to many hundreds of atomic layers of gallium and arsenic. Once the LPE growth process is in full swing, with the substrate nicely centered below the appropriate melt, the deposition proceeds at a reasonably uniform rate, so that the thickness of a particular layer can be controlled by timing how long the substrate stays in contact with the melt. The only aspects that are not so well controlled are the few tenths of a second at the beginning and end of the layer's creation during which the substrate is being dragged into and out of the melt.

As long as the fraction of time spent in those sliding start/stop manipulations is small compared to the time the substrate is left in the melt for growth, these transient effects will have little noticeable influence on the overall thickness of the layer. This condition is easily met for the thicker outer layers of the double heterostructure depicted on page 103 and, with a bit of skill and luck, is met even for that thinner central layer. But as laser structures become more complex, and layers are designed to squeeze down quantum mechanically on the electrons and holes confined within them, desired layer thicknesses begin to drop below 100 atomic layers, below the realm of thickness control in LPE. In the next chapter we will encounter such structures and the epitaxy techniques that can create them.

A close-up view reveals the sculptured "art" of the stainless steel that makes up the ultrahigh vacuum chamber of a molecular beam epitaxy (MBE) system. The small cylinder with its associated wiring seen at the lower right center is the base of one of the effusion cell ovens that vaporize molecules of source materials needed to grow modern semiconductor laser structures. The ability of MBE systems to controllably deposit single layers of atoms has led to remarkable advances in miniature lasers.

7

The Shrinking Laser

Inspired by the flexibility and variety of semiconductor materials, laser designers seek ever new ways to advance the efficiency and capabilities of their devices. An overriding theme of their endeavors has been the drive toward miniaturization. If the lasers discussed in the last chapter already seem small, just watch what is coming. Laser designers are creating smaller and smaller features, and thinner and thinner layers, all the way down to atomic dimensions.

From Quantum Wells to Quantum Dots

In the standard double heterostructure laser, the active layer can be no thinner than about 0.1 to 0.2 micrometer. Below a 0.1-micrometer thickness, light tends to diffract away, and enough of the light escapes that the electric current necessary to achieve laser threshold actually increases. The dilemma is that while it is not good to confine light in thinner active layers, it is good to confine electrons and holes, in order to build up the density of carriers that is required to reach laser threshold. In a certain sense, the dilemma tells us that electrons are smaller than photons in the visible and near visible regions of the optical spectrum that we care about here.

There is a solution to this dilemma. We can provide in effect two central regions, a thin active region for charges (electrons and holes) and a thicker confinement region for light. As long as the two regions overlap, stimulated emission can proceed efficiently. This solution has been realized in a newer and very useful laser structure called the quantum well separate confinement heterostructure, or QW-SCH.

In the QW-SCH laser, illustrated schematically on this page, the active region of electrons and holes is a very thin layer—typically a mere 10 nanometers in thickness—called a quantum well. This quantum well

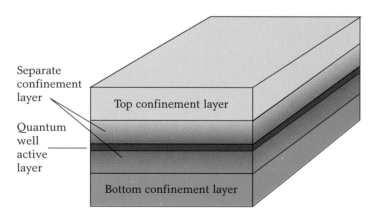

Schematic of a quantum well separate confinement heterostructure (QW-SCH) laser. The active region in which light is generated, colored red, is a quantum well (QW) only about 10 nanometers thick. The separate confinement (SC) region on either side, which traps and guides the laser light, is shown graded in color to reflect its continuous grading of composition from that of the confinement layer to one nearer that of the quantum well. The quantum well layer cannot guide the light on its own because it has a dimension much smaller than the "size," or wavelength, of the light. The outer two layers, p-type at the top and n-type at the bottom, complete the structure, making it into a laser diode.

is located in the center of a thicker confinement region for light, called the separate confinement region, which is typically about 100 nanometers thick. Electron-hole recombination in the quantum well generates light, and the separate confinement region traps it. The trapped light still overlaps with the quantum well, so the stimulated emission process remains efficient.

The quantum well is special for several reasons. First of all, it is so thin that the electrons and holes behave as charges confined in two dimensions. They are confined to a sheet, with the ability to move laterally in the plane but with no allowed vertical motion. The layer thickness is so tiny that quantum theory is required for analysis, hence the name quantum well. The quantum well laser owes its efficiency not only to the ready increase in electron and hole densities, but also to a redistribution of electron and hole states compared to the conventional three-dimensional case. In quantum wells, more electrons (or holes) are efficiently placed at the right energy to produce photons of light at the energy of the laser beam.

Only a specific number of electrons can exist at a given energy level. This statement is the extension of the Pauli exclusion principle, which for a single atom explains why only a limited number of electrons can occupy a given electron shell, to a solid material with many atoms. In the conventional case of three dimensions, the number of electron (or hole) states per unit volume is zero near the lowest available energy in the conduction (or valence) band and gradually rises as one goes to higher energies.

To build up the density of electrons needed to reach laser threshold, we must put electrons into the conduction band, first at the lower energies and then filling up states at higher and higher energies. (A similar situation applies for holes in the valence band.) The excited electrons fill states over a wide range of energies, but only a limited set of these electrons have energies at the right value for stimulated emission of photons into the laser beam. In this respect, the case of the semiconductor laser is analogous to the case of the neon gas laser, in which an incoming photon can stimulate an electron to emit a second photon only if the incoming photon energy is of a certain value. Thus we have to excite a lot of electrons to reach the population inversion condition for laser threshold, but only a limited number of those electrons can be stimulated to release the needed photons.

For the two-dimensional case of quantum wells, the distribution of electron states is significantly different. The density of electron states is a constant value from the edge of the conduction band up to some energy value, above which the density jumps to a higher fixed value. The density remains at this new value until some particular new energy value is reached and then jumps to a yet higher fixed value. When plotted, the result looks like a staircase, as shown in the center portion of the figure on the opposite page. To build up a certain density of electrons in a quantum well laser, we proceed as before by putting electrons into the conduction band, filling up states first at the lower energies and then at higher and higher energies. But because the number of states does not increase gradually as in the three-dimensional case, we can more efficiently fill in electrons for the two-dimensional case in a narrow range of energies. It is thus more efficient to reach laser threshold, and laser action begins at lower currents.

A QW-SCH laser takes less time to "warm up" compared to other lasers. When we increase the operating electric current, the density of optically effective electrons increases at a faster rate, and so stimulated emission generates photons more efficiently. This advantage is important in practice in cases where lasers are pulsed on and off rapidly to transmit bits of information, as in fiber optic communications systems. QW-SCH lasers can be turned on and off very rapidly, creating light pulses as fast as 30 gigabits per second (30,000,000,000 bits per second).

Thin quantum well active layers have one further advantage. For such thin layers, only 10 nanometers thick, the condition of lattice matching between layer and substrate is no longer so critical. A sufficiently thin layer simply expands or contracts slightly to adopt the lattice spacing of the substrate. A 10-nanometer layer is only 70 to 80 atoms deep. The atoms in a layer this thin align themselves with the atoms of the substrate and form a crystal structure in a stretched, or "strained," form if necessary. The strain alters the complex structure of the valence band in such a way that the laser threshold is still further reduced. In short, strained QW-SCH lasers offer great promise as low-threshold, high-performance lasers.

The laser becomes still more efficient if the electrons and holes can be squeezed down even more, from a two-dimensional plane to a one-dimensional line. This next step has been partially achieved in research labs, where scientists have produced the so-called quantum wire SCH laser. This type of laser is similar to the quantum well laser except that the active layer is also reduced in width to about 10 nanometers. Researchers might create the one-dimensional layer by, for example, growing the active layer in a very narrow V-shaped groove, resulting in the accumulation of a thicker "wire" of active layer material at the bottom of the groove, as illustrated on page 144. In this case the electrons are confined in both the vertical and horizontal directions. We again make use of the separate confinement region for the light.

The one-dimensional structure has a sharply peaked density of states which allows even more electrons to exist in a narrow energy range, so that more are available to emit photons at a given energy. So far,

THREE-DIMENSIONAL STATES

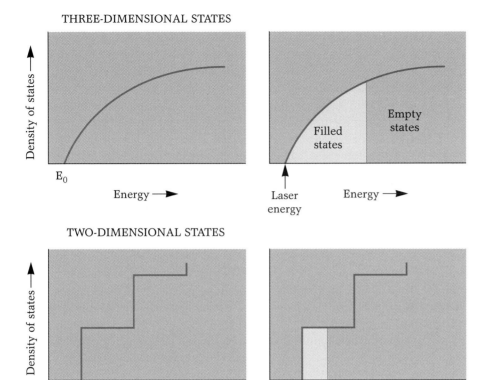

TWO-DIMENSIONAL STATES

The large number of atoms in a semiconductor provides a distribution of energy states for electrons. The number of electron states per unit volume, called the density of electron states, in the conduction band rises gradually with energy in thick active layers (top), about 100 nanometers and above in thickness. It rises in stepwise fashion with energy in thin active layers (middle), less than about 50 nanometers thick. When these structures are excited by the external current applied to the laser, the empty states fill from the lower energies upward, indicated in the right panels by the light blue shading. The carriers able to emit photons by stimulated emission reside at the energies near that indicated by the "laser energy" arrow. The fact that the density of excited carriers is higher at this energy in the two-dimensional case as compared to the three-dimensional case is the reason quantum well lasers, with their two-dimensional carrier confinement are more efficient than their conventional counterparts.

however, fabrication problems have limited the efficiency of these devices. The most severe problem is maintaining the quantum wire at a constant width along its 200-micrometer length. A typical quantum wire is only 10 nanometers wide, or 70 to 80 atoms across. Fluctuations in width along the length of a quantum wire cause severe limitations, so that, although reduced thresholds are theoretically expected, quantum wire lasers now operate at currents no lower than those that drive quantum well lasers. Researchers are optimistic, however, that with further advances in fabrication technology and materials science these devices will fulfill their promise.

The ultimate in reduced dimensionally would be a zero-dimensional device, which would confine electrons in all three dimensions. Such confined shapes, called quantum dots, would have to have sizes around $10 \times 10 \times 10$ nanometers. An electron in a quantum dot would feel extreme confinement in all directions. As a consequence, the density of electron states is a very narrow and tall spike at a given energy value in the conduction band, akin to a single electronic energy level in an isolated atom. An atom is, after all, an extreme case of confinement in all three dimensions. In such a case, all the electrons in the conduction band would have the right energy to support laser action. It has been predicted that quantum dot lasers should exhibit threshold currents that are 10 to 100 times smaller than those of quantum well lasers. Instead of requiring milliamperes, laser action would take place at 10 to 100 *micro*amperes!

Although the development of quantum dot lasers is still in its early stages, researchers have already fabricated experimental devices and operated them as lasers. In the figure on the opposite page we illustrate one type of quantum dot laser. It looks much like a conventional QW-SCH laser except that the active layer is broken up into an array of quantum dots, which can be thought of as an array of super atoms. Under the right conditions, and with the right techniques, the array of quantum dots al-

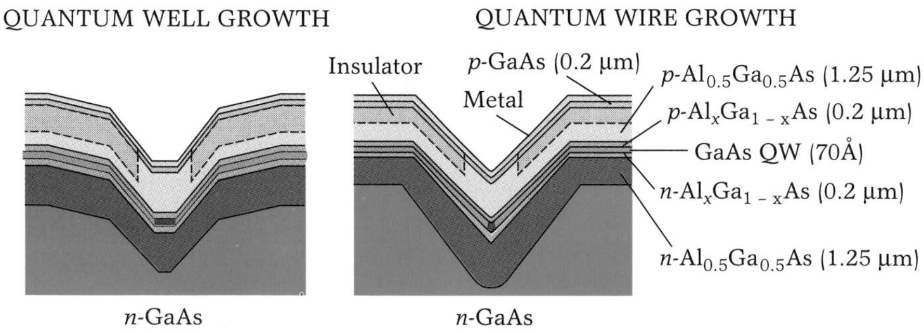

QUANTUM WELL GROWTH   QUANTUM WIRE GROWTH

Insulator p-GaAs (0.2 μm) p-Al$_{0.5}$Ga$_{0.5}$As (1.25 μm)

Metal p-Al$_x$Ga$_{1-x}$As (0.2 μm)

GaAs QW (70Å)

n-Al$_x$Ga$_{1-x}$As (0.2 μm)

n-Al$_{0.5}$Ga$_{0.5}$As (1.25 μm)

n-GaAs n-GaAs

In one scheme for building either a two-dimensional quantum well laser or a one-dimensional quantum wire laser, all layers are deposited on grooved substrates, but broad, flat-bottomed grooves are used for quantum wells (left) and sharp V-shaped grooves for quantum wires (right). For each case, laser light is generated in the base of the groove only, in a region typically 10 by 2000 nanometers at left and 10 by 10 nanometers at right.

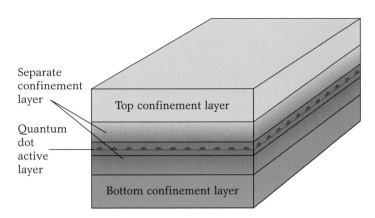

Separate confinement layer

Top confinement layer

Quantum dot active layer

Bottom confinement layer

The structure of this quantum dot laser is the same as that of the quantum well laser on page 140 except that the active region is not a continuous layer. Instead, an array of many small islands, or dots, typically 20 to 30 nanometers across and 3 to 6 nanometers high, situated in the center of the separate confinement (SC) region, forms the active regions in which laser light is generated.

most forms itself. When materials scientists deposit only 10 layers or so of atoms of certain materials, the atoms tend to bunch up naturally, like the droplets of water on a windshield, to produce arrays of quantum dots. The dots are typically about 3 to 6 nanometers in height and 20 to 30 nanometers in diameter, and rather uniformly spaced. The QD-SCH structures made to date still have high threshold currents, but progress is anticipated.

Creating layers as thin as a few atomic layers is out of the question using liquid phase epitaxy. Yet the achievement of such control is necessary not only for the quantum confinement of electrons and holes in quantum wells but also for the control of alloy composition in what are dubbed "digital alloys." In such an alloy, the proper mixture of the different constituent materials, such as aluminum arsenide and gallium arsenide, is achieved not by mixing the aluminum and gallium atoms, but by alternating ultrathin layers of GaAs and AlAs only a few atoms deep, varying the relative thicknesses of the two to vary the effective overall alloy composition. To achieve these kinds of results, we need an entirely different technique for starting and stopping layer growth, one that is controllable to even a fraction of an atomic layer. The leading process capable of such precision is molecular beam epitaxy, or MBE.

Molecular Beam Epitaxy

The MBE growth technique was developed at Bell Laboratories in the late 1960s by two men, Al Cho and John Arthur, who were not focused exclusively on seeking a new growth technique. They were initially studying the chemistry of what happens at the surface of the semiconductor gallium arsenide. To assist their studies, they had developed a way of creating an atomically clean GaAs wafer within a specially designed enclosure from which virtually all atmospheric gases had been evacuated. Having by this means preserved the clean surface of the wafer from reactions such as oxidation, they could explore how beams of gallium and arsenic reacted with the GaAs surface at different atomic fluxes, temperatures, and other conditions.

In the course of these studies, Cho and Arthur made the remarkable discovery that while for the most part the impinging gallium atoms all stuck to the surface, the arsenic atoms tended to bounce off completely when only an arsenic flux was present, but that some portion of the atoms stuck to the surface when a simultaneous gallium flux was present. Furthermore, Cho and Arthur found that the amount of arsenic that bonded to the surface was precisely equal to the amount of simultaneously arriving gallium. The perfect 50:50 mixture of gallium and arsenic was epitaxially forming more GaAs at the surface, while any excess arsenic reevaporated. Although the proper stoichiometry would have been difficult if not impossible to control manually, surface chemistry controlled it naturally. A new growth technique was born.

Crystal growth by molecular beam epitaxy is conceptually quite different from crystal growth by LPE or bulk wafer growth. MBE is more of a physical deposition process. Gaseous beams of each required atom are sprayed onto the substrate wafer in a fashion similar to spray painting. The layers deposited by MBE, as by LPE, are much thinner than the wafer onto which they are deposited, just as a layer of paint is very thin compared to the piece of wood or metal onto which it is sprayed.

All materials, when heated sufficiently, will eventually give off a vapor of their constituent molecules, just as the water in a teakettle heating on the stove gives off steam. Our everyday experience with a material such as aluminum tells us that it is a solid metal, but heated above 660 to 1200 degrees Celsius it too becomes a molten liquid emitting a continuous stream of vapor, in this case formed of aluminum atoms. To hold the molten aluminum, we construct a container, referred to as a "crucible," with a single open hole at one end. The stream of aluminum atoms escaping from the surface of the melt will exit the container in a single direction, through the hole, to form a beam, just as the teakettle's spout sends the steam out in a particular direction.

Building upon Cho and Arthur's original finding, we first create a beam of arsenic that continuously bathes the substrate. Without a gallium beam no crystal growth occurs. Turning on a gallium beam by opening its individual mechanical shutter begins the creation of a gallium arsenide layer. Should the next desired layer be, say, aluminum arsenide, a mechanical shutter leading into the path of the gallium beam can be closed, and one in the path of the aluminum beam opened, to shut off the gallium beam and release the aluminum beam. If we want to create

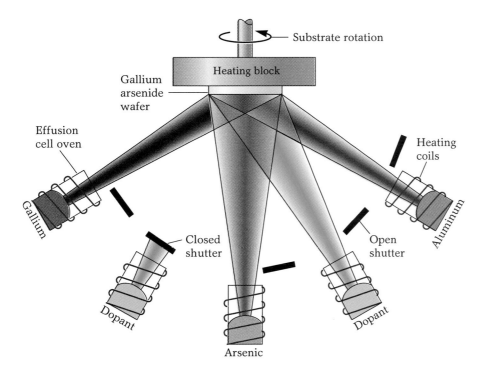

A schematic of a molecular beam epitaxy (MBE) system, showing five ovens that vaporize and direct "beams" of atoms or molecules through a high-vacuum space, converging onto a substrate wafer. The substrate, shown here as a gallium arsenide wafer, is mounted on a heated block and rotated continuously to promote uniform crystal growth on its surface. Here the shutters are adjusted so that beams of the right-hand dopant together with gallium, aluminum, and arsenic impinge simultaneously on the substrate, creating a layer of doped aluminum gallium arsenide. The left-hand dopant's shutter is closed, preventing its incorporation into the layer.

an AlGaAs layer, we open the shutters for both the aluminum and gallium beams, creating the necessary mixture of aluminum and gallium atoms in the growing layer just as the simultaneous spray painting of red and white paint produces a layer of pink. When we are finished growing our crystal, we simply close both the aluminum and gallium shutters.

In our spray paint analogy, a fast white spray being codeposited with a slow red spray would lead to a very light pink shade. In the same way, we can control the fraction of aluminum, x, in our $Al_xGa_{1-x}As$ alloy by properly regulating the flow of atoms in the aluminum and gallium beams. Thus the one final ingredient we need is a way to control the rate at which the atoms spray in each molecular beam, as measured by the number of layers of atoms arriving per unit time.

For our teakettle, we know all we need to do to make the stream of steam stronger is turn up the heat. Such is true for any of the elements. We rely on the fact that the hotter the element gets, the more atoms escape from its surface to enter the gas phase. Scientists use the concept of vapor pressure to quantify this. Suppose that an evaporating solid or liquid is placed in a closed container from which the released gas-phase atoms cannot escape. The atoms escaping from such solids or liquids (from what are termed condensed phases) rattle around inside the

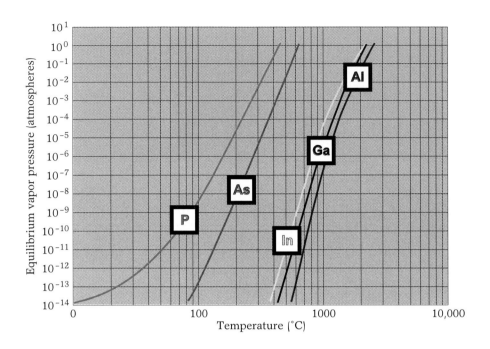

The equilibrium vapor pressure plotted against temperature for five elements used in MBE gives a relative indication of how rapidly each element will evaporate out of its oven, or "effusion cell," at any given temperature. The graph shows that phosphorus and arsenic can be vaporized at lower oven temperatures than the metals indium, gallium, and aluminum. The magnitude of the vapor pressure, and hence the evaporation rate, can be adjusted by many orders of magnitude through relatively small changes in oven temperatures.

container, bouncing off the heated walls, until they eventually hit the condensed-phase solid or liquid from which they originally came and re-condense. The graph on this page indicates for some of the elements used in MBE the pressure known as the "equilibrium vapor pressure," exerted by the gas vapor in such an enclosure when there are just enough atoms in the gas phase so that the flux of gaseous atoms return-ing to the condensed phase balances the flux of new atoms escaping from it. The equilibrium vapor pressure of a material is a direct indica-tion of how quickly atoms will escape from it by evaporation. We can use these universally determined values of the equilibrium vapor pres-sure of the elements to give us the necessary temperature dependencies of our "teakettle" sources. It can be seen in the figure above that the tem-peratures needed to attain a given pressure, and hence a given evapora-tion rate, vary widely from one element to the other, so the tempera-tures that the various atomic "teakettles" must be maintained at depend strongly on the element. Furthermore, the vapor pressure and hence the evaporation rate can be conveniently varied over many orders of magni-tude by simply carefully controlling the temperature of the crucible.

So the crystal grower's "teakettle," or "effusion cell" in the official parlance, consists of an open-ended crucible containing the solid or liq-uid form of an element, heating coils and heat-reflecting baffles around the crucible to elevate it to the appropriate temperature, and a temperature-sensing element, usually a thermocouple, to measure the exact cell temperature. The photo on this page shows three such effusion cells within an MBE chamber. The crucibles containing the heated elements have been removed to reveal the heating coils glowing red-hot.

As Cho and Arthur discovered at the outset, the MBE process needs to take place in an ultraclean environment. Not only does the chamber in which the process goes forward need to be dust free, it needs to be as free as possible of the gases ordinarily present in air. Otherwise, the substrate is exposed to a constant flux of unwanted impurities as the air's gas molecules strike the wafer's surface. To avoid such impurities, the process is carried out in a vacuum enclosure from which virtually all atmospheric gases have been pumped out. State-of-the-art MBE systems achieve "ultrahigh vacuum" (UHV) levels so high that the residual gas in the chamber is present at a gas pressure roughly 10,000,000,000,000

Three effusion cell ovens glow red-hot inside an MBE system as they point upward toward a substrate out of view beyond the top of the photo. The cups, or "crucibles," containing the elements to evaporate have been removed in order to reveal the complex structure of the heating filaments. These filaments have been designed to give steady, uniform heating to the crucibles and the source material within them.

Together with a colleague on the other side of a molecular beam epitaxy system, one of the authors (James Harbison) adjusts the position of the substrate with the knob in his left hand while he examines the pattern on the electron diffraction screen used to monitor the growth of the layers in real time. A myriad of wires are connected to the MBE system to control and monitor crucial parameters such as oven temperatures, substrate temperature, and pressure. The thicker black tubes in the foreground deliver a steady stream of liquid nitrogen used as one of the means to pump the vacuum of the system by freezing any remaining gas onto cooled panels inside the stainless steel growth chamber.

(10^{13}) times smaller than normal atmospheric pressure! Pressures in the molecular beams themselves can be as high as a million times this "base" pressure of the chamber, but they cannot be much higher or atoms within the beams will begin to bump into one another and scatter off in odd directions out of the beam. At this level of vacuum, among the highest achievable in the laboratory for a vessel this size, it is possible to limit the unintentional doping of the crystal to one part per ten million or lower.

The photo on this page shows the outside of a typical MBE system. Most of the complexity in its design, as well as most of its cost (on the order of a million dollars!), comes from the constraints of working at UHV vacuum levels. This complexity does lead to a rather visually appealing piece of equipment, and its glamour has had a number of interesting side effects for one of the authors, James Harbison, as an MBE grower. It becomes one of the prime subjects for a high-tech photo, such as the one on the previous page, which gives the operator high visibility but sometimes makes it hard to get work done! The fancy equipment also makes the job look appealing to others. When asked after visiting his lab what it is that his dad does, the author's four-year-old son explained with approving wide eyes, "He pushes buttons all day!"

So given this specialized environment, how exactly does the growth itself occur? The crystal grower begins with a wafer substrate of single-crystal structure to serve as the template, no different from the crystal substrate used in liquid phase epitaxy, though it is larger, several centimeters across. Once the wafer has been introduced into the vacuum through a load-lock arrangement, akin in many ways to the load lock the astronauts use to leave the space shuttle and enter the surrounding vacuum of outer space, it is positioned at the point in the chamber where all the beams converge, as shown on page 147.

Before the wafer can be exposed to any molecular beams, it must be heated to the proper temperature. Remember that in LPE the choice of substrate temperature is severely constrained. The temperature of the substrate in LPE and the melt with which it is in contact has to be very close to the actual freezing point of the melt. If it is too low, the under-cooled melt will grow too quickly onto the substrate, making thickness hard to control. Worse, some of the atoms being incorporated into the growing solid may lack the time to find their ideal crystalline positions, causing uneven or defective crystal growth. If the temperature is significantly above the freezing point of the melt, molten material will not deposit onto the substrate. In fact, the substrate itself may begin to dissolve! In MBE, there is no contact between the substrate and the molten elemental material in the crucibles. Thus the atoms in the crucible material and the atoms in the substrate need not be in a delicate equilibrium as is required in the LPE case. In fact, they are very far from equilibrium. The flux is almost entirely one way, from the cells to the substrate. There is virtually no flux of atoms from the substrate back into the cells.

So we are free to choose a substrate temperature independent of the cell temperatures, which, you will recall, have been independently chosen to provide the proper atomic flux for each molecular beam. Two factors limit how high the substrate temperature can go. The temperature should not be so high that the gallium atoms impinging on the substrate can easily reevaporate, thus making control of the net overall deposition rate and, in the presence of another Group III flux such as aluminum (which reevaporates more slowly), even the relative composition of the film, much more complex and sensitive to substrate temperature. In addition, the atoms in the already deposited layers must not become so hot

Atoms arrive (1) from a "molecular beam" as they reach the substrate surface. The atoms stick to the surface and migrate around (2) until they find a site (3) where sufficiently strong bonding occurs. That site is typically at the edge of a spreading atomic layer, the growing epitaxial crystal. The newly forming layer is shown in a different color for the sake of clarity. Group III atoms arrive at the surface as shown as single atoms, though the case is slightly more complex for Group V beams, which consist of molecules such as As_2, As_4, P_2 or P_4. This is the origin of the term "molecular beam."

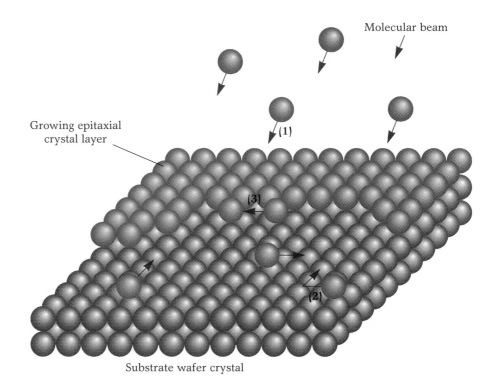

that they have enough energy to hop around, or, in the language of the materials scientist, "diffuse," within the crystal itself and blur the boundaries between the distinct layers.

Both these factors would argue for growing the crystal at lower temperatures. Why then elevate the temperature of the substrate at all? This question takes us to the heart of the MBE growth process—the way it proceeds, at the fundamental atomic level, depicted schematically in the figure on this page. Each molecular beam must provide just the right number of atoms needed for the growth of the current layer, shown in red in the figure. We can control their arrival rate at the surface (1) using knobs that regulate the cell temperatures and shutters, but we can't control their subsequent incorporation into the proper atomic sites, the actual crystal growth itself. This nature does for us. The atoms arrive at the surface of the substrate and, aided by the motion imparted to them by their thermal energy, begin to migrate around that surface (2), seeking a proper binding site.

The surface atoms of the substrate have bonds sticking out like arms in prescribed directions, reaching to join with neighboring atoms. We can think of this surface migration as the process of finding the right "set of open arms." When an atom is still out on the open "blue plain" of the surface, as in (2), it can bond only with the single atom directly beneath it. However, if we have adjusted our substrate temperature to be high enough, the roaming atoms possess enough thermal energy to continually shake loose from these single connections, and they continue to travel on across the surface. When, however, a migrating atom reaches the edge of the newly forming layer, as in (3), it can form bonds to atoms to the side as well as below, and it becomes more tightly linked. The atom ceases its movement across the surface and drops into place in its proper position in the growing crystal. In such a way, the crystal grows an atom at a time. This process fixes another set of bounds on the substrate temperature. The temperature should be hot enough that thermal energy shakes the atoms loose when they're on the open plain, but not so hot that they will be dislodged from a proper binding place at the step edge of the growing layer.

The atoms arriving at the surface must be allowed sufficient time to reach their proper position at the step edge before an entire new layer comes down and buries them. Otherwise, we will get a very rough surface with all kinds of mountains and valleys on it. Worse yet, the crystal can actually end up with defects, such as missing atoms at sites in the crystal structure, that result in undesirable electrical properties.

We must also take into account another practical constraint on the growth rate of the crystal: grow the crystal too quickly and we lose the ability to controllably fabricate layers as thin as one or two atoms. We start and stop layer growth by opening and closing mechanical shutters, and these take a fraction of a second to entirely open or entirely close. So the time it takes to deposit a layer must be considerably longer than the time it takes to open and close the shutter. If we deposit at a rate such that a new atomic layer is constructed about once a second, then we will, in principle, be able to construct crystals with epitaxial layers only a single atom thick! In a sense, such precision can give us ultimate control of the crystal, at least in the vertical direction. We adjust the arrival rate of the atoms to achieve this desired single layer, or "monolayer," per second growth rate, then we heat the substrate just enough to allow those atoms wandering on the surface to make it to the edge in time.

But we cannot adjust the arrival rates of the atoms unless we know what those rates are. So the final question is how we know in fact what those arrival rates are and, even more important, whether the atoms have enough "mobility" on the surface to make it to the growing edge. It is in exploring the answers to these questions that a crystal growth technique like MBE really shines, because the ultrahigh vacuum environment necessary for clean crystal growth is also the ideal environment for probing the inner workings of the atomic processes at the surface itself.

The OMCVD Method of Crystal Growth

In addition to liquid phase epitaxy (LPE), now largely passé, and the more modern molecular beam epitaxy (MBE), there is another important method for preparing semiconductor laser materials. The method is called organometallic chemical vapor deposition, or OMCVD for short. Like MBE, the method delivers materials to the growing layers in the gas state. Unlike MBE, however, the gases are not atoms in elemental form sprayed directly from source to crystal, but chemical compounds delivered through a set of pipes and valves. These compounds are combinations of organic molecules and metals—hence the use of the term "organometallic." An example is trimethyl gallium, a combination of the organic molecule methane (the main constituent of natural gas) and the metal gallium. It is the metal part of the compound that will be incorporated into the growing crystal layer. These types of compounds can deliver the metal conveniently to the desired site because they stay in a gas state until they thermally break up in the vicinity of a heated substrate. The metal atom sticks to the substrate, combining with other atoms, while the organic part of the compound remains as a residual gas (methane in the case of trimethyl gallium) that flows away cleanly. The

A crystal grower holds a 2-inch-diameter indium phosphide substrate, just removed from an OMCVD system's quartz reactor, upon which he has grown a laser structure.

Monitoring Growth an Atomic Layer at a Time

The tool MBE crystal growers use to explore these microscopic aspects of the growth process is known as reflection high-energy electron diffraction, or RHEED. The figure on page 156 shows the basic components of a RHEED system. One crucial component is an "electron gun" con-

OMCVD method can efficiently deliver atoms of just the elements needed for modern semiconductor lasers, including indium, gallium, arsenic, and phosphorus.

Crystal growth takes place within a reactor like that seen in the photograph in this box. Inside the clear quartz reactor tube, a graphite holder carries a standard semiconductor substrate; the tube is surrounded by the metal coil seen in the photo, which provides heating capabilities somewhat like those of a microwave oven. Various gases flow from tanks through a manifold of pipes and valves that can be used to select different gas mixtures. The selected mixture flows into the reactor from the left side of the photograph, and the gases deposit the required semiconductor materials onto the substrate as epitaxial layers to form laser structures. The system easily switches from one set of gases to another to grow epitaxial layers of different composition.

Because the rate of deposition in OMCVD is governed by chemical reactions at the heated substrate, layer thicknesses and composition depend critically on a number of changeable parameters such as substrate temperature, gas pressure, and gas flow rates. Thus users of OMCVD must perform extensive calibrations on test samples in order to get these parameters properly set. In marked contrast, the purely physical deposition process of MBE, because of its straightforward simplicity, lends itself to the rapid creation of entirely new types of structures. Once properly calibrated, however, the OMCVD method can produce high-quality semiconductor crystals about 10 times faster than MBE. Another advantage is the ease with which gas sources may be switched, just by changing tanks of gases connected to the manifold. The process is conveniently controlled by a computer that automatically opens and closes valves to allow appropriate gases to enter the growth chamber. Its speed and flexibility make the OMCVD method attractive for both commercial growth of laser structures and laboratory use.

The OMCVD and MBE methods are both important for the growth of compound semiconductor crystals. Each has its pros and cons for specific applications. There is even a technique, called OM-MBE, in which some of the solid materials that serve as sources in an MBE system are replaced with connections to some of the pipes from an organometallic gas source, effectively combining the two techniques.

taining a hot filament that serves as a source of electrons. A series of electric coils and plates within the gun serve to accelerate the electrons and direct them into a tightly focused beam. In fact, this electron gun is identical to the device used in a television tube to create the flickering beam that excites the glowing phosphors on the front of the set. A close colleague of one of the authors, lacking the ten or twenty thousand

The reflection high-energy electron diffraction (RHEED) system used in MBE to monitor crystal growth. The electron gun generates a beam of electrons (red line) that glances off the growing crystal surface and scatters toward the phosphorescent screen. The screen lights up where electrons strike it, revealing a diffraction pattern as well as a specular spot reflected directly off the surface as from a mirror. A photodetector placed next to the specular spot measures its varying light intensity and produces the chart shown, in which a single up-and-down trace indicates the growth of one complete atomic layer.

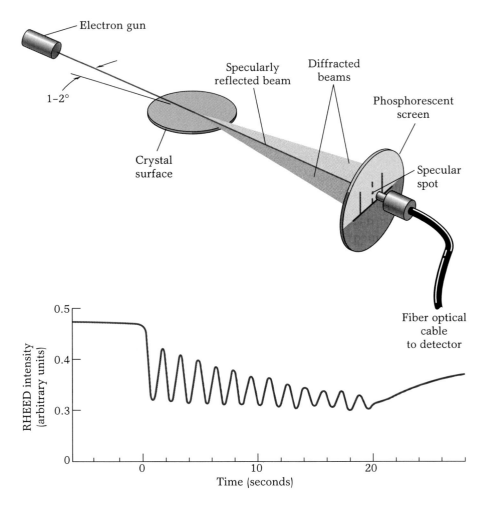

dollars necessary to purchase such a sophisticated RHEED gun for his MBE system, instead converted parts from an abandoned TV set into just such a gun, and it worked quite acceptably! Opposite the gun is a phosphorescent screen coated with a phosphor designed to glow when it is exposed to the beam of electrons. The gun and the screen are placed at opposite sides of the chamber, so that when the electron beam is aimed straight at the screen one sees a single bright focused spot on it.

The system becomes much more interesting when the substrate is moved into the beam. The electrons, instead of hitting the screen directly, glance off the crystal and reflect onto the screen. A pattern is formed on the screen, known as the "diffraction pattern" (hence the term "diffraction" in the RHEED acronym), and it reveals a great deal about the arrangement and spacing of the atoms at the crystal surface, in much the same way that X-ray diffraction reveals the crystal structure of bulk crystalline materials. But though this entire diffraction pattern contains a wealth of information about the crystal structure, it is the single spot in the pattern that represents the direct reflection of the tightly focused beam itself, reflected as a beam of light would be reflected in a mirror, that reveals how rapidly the atoms are arriving and how effectively they are moving about on the surface.

This spot on the screen is referred to as the "specular spot," since it is created only by electrons that are directly, or "specularly," reflected off the surface without being diffracted by the underlying crystal structure itself. Think of it as analogous to a narrowly focused flashlight beam bouncing at an angle off the surface of a polished piece of metal. When the surface is mirror-smooth, the beam is reflected entirely intact, and if we place a light detector on the wall opposite the light where the reflected beam ends up, we get a good high light reading. If, on the other hand, we were to roughen the metal's surface, some of the flashlight's beam would be scattered off at random angles, leaving less light falling into the light detector. The rougher the surface, the less light makes it to the detector. So the intensity of the beam in the detector is a direct measure of the smoothness of the macroscopic metal surface.

With RHEED the same process occurs, only now we measure the intensity of a reflected electron beam, one with an electron wavelength small enough to be sensitive to the roughness of a surface on an atomic scale. We still use the intensity to infer the overall roughness, but this time it is the atomic roughness of the crystal's growth surface. This roughness may be so small in scale that it cannot be seen even with the world's most powerful microscopes, yet we can get a direct measure of its magnitude by observing the intensity of the reflected electron beam, without ever seeing the hills and valleys that make up the roughness.

The figure on the next page shows what we see during crystal growth when we look at surface roughness by means of this RHEED tool. Initially the surface is smooth, and the specular electron reflection is at its maximum intensity (1). Then, as we begin to add atoms to the next layer, it becomes increasingly rough, as islands of atoms begin to form in the new layer (2). This roughness reaches a maximum at about half a layer of coverage (3), but thereafter the surface becomes smoother and smoother as the islands coalesce (4). Finally, the surface returns to its original smooth state once the layer is completed (5). The time trace of the specularly reflected electron beam, indicated in the right-hand side of the figure, shows a complete cycle. If we were to continue this process for many layers, we would observe the series of regular oscillations shown at the bottom of the figure on the facing page, referred to by MBE crystal growers as "RHEED oscillations."

The steps in the MBE growth of one epitaxial layer of atoms. The sequence at the left shows how atoms in the next forming layer (red) first form islands on the substrate surface (blue) that gradually grow out into a complete atomic layer. The rougher the surface, the less the electrons are reflected off it specularly, and the lower the intensity of the specular spot. The series of right-hand panels indicate the intensity of the specular spot in the RHEED pattern at each stage of growth. The specular spot goes through one complete cycle from bright to dim to bright as the atomic layer grows from start to completion.

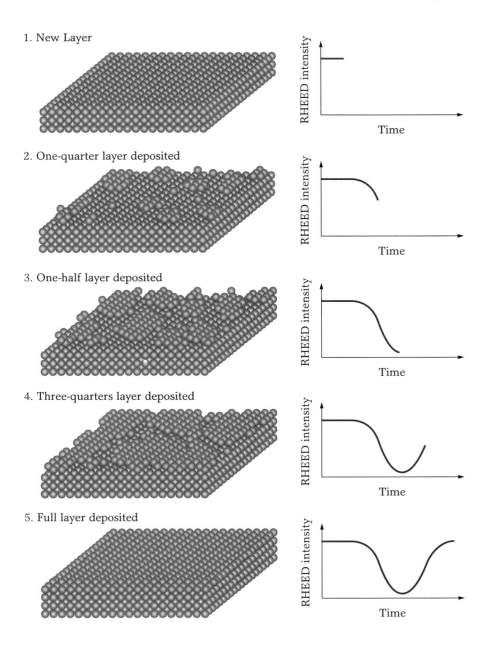

1. New Layer

2. One-quarter layer deposited

3. One-half layer deposited

4. Three-quarters layer deposited

5. Full layer deposited

To see this RHEED oscillation process in action can be awe inspiring. You open a mechanical shutter, and with your naked eye you can see a tiny spot on the phosphorescent screen winking at you, counting the addition of each layer of atoms, one by one, before your very eyes. You're watching the atoms go down a layer at a time! Furthermore, with a ten-dollar stopwatch and a bit of patience you can measure the rate of crystal growth accurately to within about 1 percent, even though the crystal is growing at the rate of only about a thousandth of a millimeter an hour. Although the author writing these words has witnessed this process thousands and thousands of times, it still evokes a sense of wonder in me. The crystal grower has truly approached the point of being able to lay down the desired epitaxial layers with literally atomic layer precision, and with RHEED oscillations is able to actually observe the process happening in real time.

It now becomes obvious how the MBE crystal grower can determine the rate at which the crystal is growing. As long as the grower has in mind the well-known atomic layer spacing of the particular crystal being grown, the rate of the RHEED oscillations can give a direct measure of the deposition rate. Creating layers of precise thicknesses, even when those thicknesses are as small as a few atomic layers, becomes a matter of simply timing the opening or closing of mechanical shutters. This step can be performed manually, as one of us (James Harbison) often does when I need to grow a new structure just to the point where I see the proper structure in the RHEED pattern, or, more commonly, a computer can control the shutter timing. Furthermore, before beginning to construct a crystal, the grower can precisely measure the rate at which atoms arrive at a test substrate for each of the needed molecular beams one at a time, then achieve the proper rates by adjusting the cell temperatures, in this way, together with shuttering, controlling both the layer thicknesses and the precise fractions of atoms in alloy mixtures. This powerful tool can even answer the question of how hot the substrate needs to be to allow atoms ample time to migrate to step edges—the crystal grower need only observe the temperature range over which the RHEED oscillations occur. At too low a substrate temperature the atoms can't reach the growing edges in time before the next layer arrives. The regular atomic layer-by-layer progression breaks down, and the RHEED oscillations disappear. In sum, these oscillations give the grower a unique insight into the actual growth process itself.

With the power of a modern technique such as molecular beam epitaxy, today's laser designers are able to devise layer structures that customize properties such as the crystal's optical emission wavelength, dopant type, or electrical conductivity in an almost arbitrary way in the vertical direction of growth. This ability, applied to the construction of a simple double heterostructure laser, can lead to a compact, efficient laser the size of a grain of salt. It has, as well, made possible the design of a wide diversity of new semiconductor laser structures, culminating in the creation of a laser so small that a million will fit on a piece of a wafer the size of a thumbnail.

The Tiniest Lasers

Most of the semiconductor lasers we have discussed thus far have, as their basic geometry, the structure shown on the left in the figure on the facing page: the light emerges from the edges through mechanically cleaved mirrors, formed by cleaving the semiconductor crystal along a direction perpendicular to the central glowing active layer. These "mirrors" are reflecting, but certainly not perfectly so. They are more akin to a window pane, which reflects some light back but also allows a great deal to pass directly through. This phenomenon of partial reflection occurs whenever light passes from one material to the next if the two materials have a different index of refraction, a measure of the degree to which the material interacts with a light beam traveling through it.

The cleaved semiconductor facet allows out about 70 percent of the light into what becomes the external laser beam, and reflects about 30 percent of the light back into the glowing active layer, enough to provide a supply of photons to stimulate more emission. During that return trip, before reaching the second mirror, these photons need to stimulate enough additional light to make up for the 70 percent lost out the first mirror. This requirement limits how short the laser can be.

If a significantly smaller fraction of light leaks out the mirror, then the returning photons need traverse significantly less material to rebuild their numbers. One way to reduce the leakage a bit is to coat the cleaved facets with extra layers of optical material, usually clear insulating compounds (consisting of oxides or nitrides generally referred to as dielectric materials), of alternating index of refraction. Some additional fraction of light will be reflected at each successive interface, thus recovering a bit more light before it escapes out the end edge of the laser. Through such an arrangement, the mirror can be made to reflect a higher percentage of the internal light beam, but the returning photons still need to make up for the percentage lost out the end. With such an arrangement the active layer can now be as short as about a tenth of a millimeter, or about one hundred micrometers, but no shorter.

This size, on the order of the thickness of a human hair, is certainly small compared to the size of a normal gas laser, but it is rather large compared to the features on a standard silicon integrated circuit, which are only a few micrometers or less on a side. One of the challenges faced by the laser design community at the end of the 1980s was how to drop the size of a semiconductor laser down to a few micrometers.

Such a feat would allow the fabrication of many more laser devices on a given chip, thus potentially driving the cost down precipitously. Moreover, these tiny structures would require significantly less current to lase, since a much smaller volume of material in the active layer

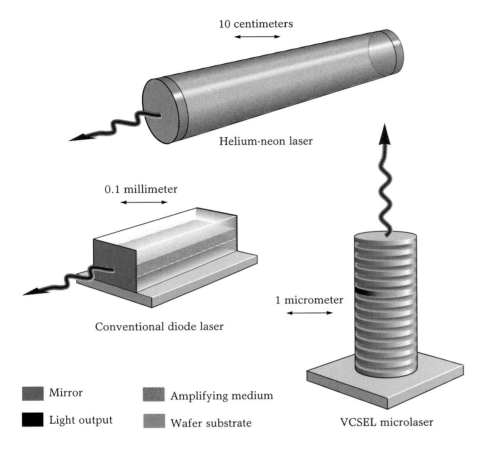

10 centimeters

Helium-neon laser

0.1 millimeter

Conventional diode laser

1 micrometer

VCSEL microlaser

Mirror

Light output

Amplifying medium

Wafer substrate

A comparison of three laser geometries: the gas laser, the conventional semiconductor diode laser, and the new vertical-cavity surface-emitting laser (VCSEL). All three contain an amplifying medium and a pair of mirrors, although the type of materials employed and the geometries differ markedly. Note the different length scales, ranging from relatively large, to small, to extremely tiny.

would have to be electrically driven, or in laser parlance "pumped," to glow. The answer clearly lay in increasing the mirror reflectivity. The question remained, how to increase this reflectivity significantly.

In the last decade, as often happens in science, a number of crucial experimental advances, achieved in a variety of research areas, together provided the key that unlocked the solution to this technological problem. A group of researchers in Japan led by Ken Iga at the Tokyo Institute of Technology had been pursuing the novel idea of tipping the semiconductor laser on its side so that the photons would reflect back and forth along an axis perpendicular to the deposited epitaxial layers. The distance traveled through the glowing active layer thus became the *thickness* of the layer, which since it was grown epitaxially could be at most a number of micrometers. Such a "vertical-cavity" laser unmistakably required extremely reflective mirrors.

The group managed to deposit such mirrors in the form of alternating dielectrics of differing indices of refraction, quite similar to the highly reflective coatings already being used on the cleaved facets of edge-emitting lasers. The dielectric layers were deposited both on top of and beneath the active layer, and though they worked reasonably well as optical reflectors, such dielectrics are electrically insulating and therefore made the injection of electrons and holes from above and below into the active layer quite difficult. Consequently, complex schemes had to be devised to get the electrons and holes, the charge "carriers," into the active layer of this "sandwich" structure from the side.

After prolonged effort the group was able to achieve lasing in such structures, but only by applying an amount of current that was far from being desirably low in comparison to that routinely attained for edge emitters. These early vertical-cavity lasers were in fact such great electricity guzzlers, and achieved lasing at the cost of such extra complexity, that they could be regarded only as an interesting demonstration of principle, still clearly out of the mainstream. But they showed the promise of a new kind of laser.

A second experimental thread being developed in this time frame was the ever increasing precision of the layer thickness attainable by advanced crystal growth techniques. In particular, RHEED oscillations could accurately monitor deposition rates in molecular beam epitaxial growth to a precision on the order of about 1 percent, a dramatic improvement over techniques of the past. A group of researchers led by Jack Jewell at AT&T Bell Laboratories exploited this precision to investigate optical effects in vertical cavities that were not lasers. Jewell's group tested a structure consisting of a central layer surrounded by stacks of alternating GaAs and AlAs layers constructed to act as mirrors, analogously to highly reflective dielectric mirrors. The scientists made their mirror layers of a thickness chosen so that the crests and troughs of the partially reflected waves from all the interfaces would be "constructively" in phase, and thus sum together, as they returned back into the cavity. As seen in the figure on this page, such constructive addition is possible for only the single specific wavelength for which the round-trip distance to the next set of layers and back is one full cycle. The partial reflections of other wavelengths of light will not match in this way, and these reflections will partially cancel one another, leading to a relatively low reflectivity. The same concept had been employed in making the high-reflectivity coatings used on the shortest edge-emitting laser mirrors and Iga's vertical-cavity lasers, but now researchers further refined the technique by constructing the alternating layers out of semiconductors over which the crystal growers had gained an unparalleled degree of control. Using RHEED, the mirror layer thicknesses could be very precisely chosen to highly reflect the particular wavelength of interest. In addition, the layer thicknesses could be grown uniformly across an entire substrate wafer, so that the optimal reflection wavelength was the

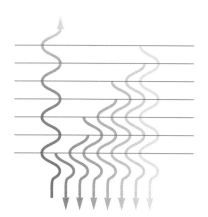

Alternating layers of GaAs and AlAs materials provide constructive reflectance for a specific wavelength. The upward-heading wave, left, has just the right wavelength so that it is partially reflected at each one of the interfaces, producing downward-heading waves from the interfaces between layers that all add together in phase.

same for the devices across the whole wafer. The result was a vertical cavity with mirror stacks attaining reflectivities as high as 99.9 percent! This encouraging result was even more exciting when considered in light of the vertical-cavity laser geometry being studied in Japan.

A third crucial experimental achievement reached fruition in the laboratory in this same time frame, and it was again related to the U.S. vertical-cavity optical experiments. This was the development of highly directional etching techniques that could be used to cut down into semiconductor materials, leaving nearly vertical sidewalls. Of particular importance was the ability, newly discovered by Bellcore researcher Axel Scherer, to perform this feat on even a complex layered structure such as the new mirror stacks. These stacks are made up of a number of chemically distinct materials that should chemically etch in quite different ways. The goal was to pattern the top of the stack with a protective mask, say in the shape of a disk, then to etch down vertically, removing material everywhere but beneath the mask to leave a freestanding cylinder.

A stream of ions impinging vertically on the surface could physically erode the material not protected by the mask, but this technique, although it produced admirable results, etched too slowly. An alternative technique was to expose the semiconductor to a gas that reacts with the surface. This technique is much faster than such ion beam erosion, but the pattern of etching cannot be controlled, since the reactive gas can etch in from the side beneath any mask, and the technique etches different materials at different rates.

The solution, developed in the 1980s and referred to as chemically assisted ion beam etching (CAIBE), is to expose the layer surface to a chemically active gas that reacts with the semiconductor, but whose reaction products fall just shy of having the necessary vapor pressure to leave as a gas. A collimated ion beam impinging on the surface activates these reaction products—tickles them in a sense—and releases them into the gas to remove them from the semiconductor, thus exposing new semiconductor surface for further etching. Surfaces exposed to the ion beam etch very rapidly, whereas those shielded from it by the mask, including the vertical sidewalls, etch hardly at all. Thus CAIBE achieves the best of both techniques: like ion beam etching, it creates highly vertical sidewalls but performs this feat at the rapid rates usually only attained by chemical methods. This third breakthrough allowed the fabrication of vertical-cavity optical resonators, quite akin to a vertical-cavity laser, entirely fabricated out of semiconductor. All that had to be added was a layer in the center of the cavity containing excited atoms ready to be stimulated to emit light.

Here one more obstacle presented itself. If the mirror was to be made of GaAs and AlAs layers, the laser designer would have to make the light-emitting region out of a material with an energy gap smaller than that of both these materials since, as the reader will recall from

An electron microscope image, viewed from above, of an array of VCSELs (vertical-cavity surface-emitting lasers) ranging in diameter from 1 to 5 micrometers. All were produced on the wafer during a single etching process, capable of forming literally millions of lasers at a time.

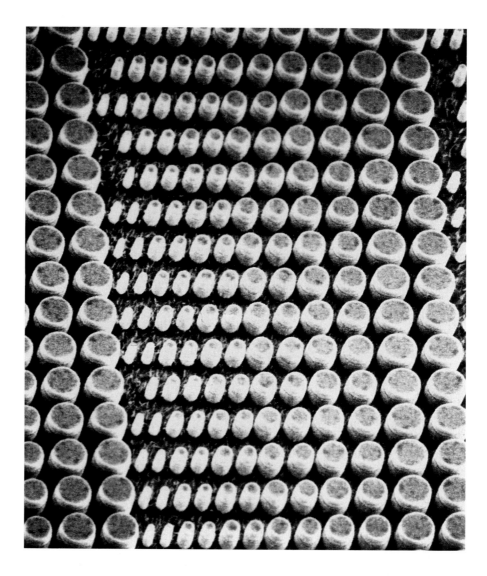

Chapter 2, materials are only transparent for light whose energy is lower than their particular energy gap. There are no materials with a bandgap smaller than GaAs that are also lattice matched to it. Here a final advance achieved in the 1980s again came to the rescue. Semiconductor crystal growers had discovered that, under the proper conditions, a material with a lattice constant different from that of the substrate could still grow perfectly without defects below a certain critical thickness. The secret, as discussed earlier, was to allow the crystal lattice of the growing layer to be strained, to contract or expand to match that of the substrate. Thus a vertical-cavity laser with GaAs/AlAs mirrors could be fabricated by choosing the smaller bandgap material indium gallium arsenide (InGaAs), which has a slightly larger lattice constant than GaAs, as the active layer between the mirrors and allowing it to grow strained while keeping its layer thickness below the critical thickness.

With the pieces all in place, one of the authors, James Harbison, and his colleague at Bellcore Leigh Florez grew such a structure to Jack Jewell's design specifications, using MBE to form both the mirrors and the strained InGaAs active layer material between them, while Axel Scherer applied his newly refined CAIBE etching technique to fabricate the freestanding laser structures. To their great surprise, the very first devices succeeded in lasing, and within months devices were being made that rivaled the best edge-emitting lasers in their low usage of electric current.

An electron microscope image of an array of VCSELs viewed from the side. The materials GaAs and AlAs etch at slightly different rates, producing the lateral patterning seen on the outer walls of the upper and lower reflector layers. This patterning clearly reveals the layered structure of the lasers' mirrors. The central, relatively thick region is the active region containing some indium gallium arsenide quantum wells.

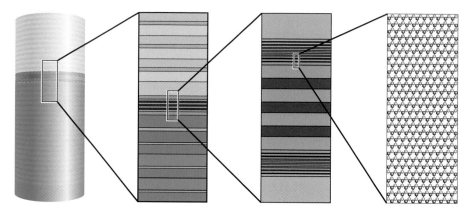

The anatomy of a VCSEL at various levels of magnification. The full structure seen at the left shows the GaAs-AlAs mirror sections (*p*-type at the top in light green and blue, *n*-type at the bottom in deep green and blue) and the indium gallium arsenide active layer in the middle (red). The laser beam emerges downward from the bottom end of the structure. Sequentially higher magnifications are shown from left to right, revealing the detailed layering of materials in the laser, which contains about 600 layers in all. At the far right the magnification is so large that individual atoms can be observed, aluminum atoms depicted in blue, gallium in green, and arsenic in gray. Notice in the second panel from the right how an effective gradation from aluminum gallium arsenide, shown in blue-green, to gallium arsenide in green, above and below the three red quantum wells in the central active region, is achieved by cycling back and forth between the two while varying the proportion of each in successive cycles. This technique is referred to as "digital alloying."

Specialists refer to these microlasers as vertical-cavity surface-emitting lasers (VCSELs), because their light emerges out of the wafer's surface rather than edge. Edge-emitting structures need to be precisely broken, or "cleaved," individually to form each device, whereas in the case of these vertical-cavity microlasers a single etching step forms literally millions of lasers simultaneously. The ease of mass manufacture is one of the advantages of creating microlasers by an etching process like CAIBE. The reasonably high cost of the etching fabrication step becomes distributed across these millions of individual devices, so that the cost on a per-laser basis becomes vanishingly small. It is precisely the same phenomenon that allows the fabrication of a computer chip containing literally millions of transistors at a cost of only a hundredth of a cent per device! When one thinks of how the tremendous drop in price per transistor has affected the world of electronic integrated circuits, making possible an array of electronic devices ranging from digital wristwatches to laptop computers to cellular phones, one is tantalized by the possibilities such parallel fabrication of semiconductor lasers might bring.

Let's look more closely at one of these VCSELs. The figure on the facing page sketches the anatomy of an individual surface-emitting microlaser at ever increasing levels of magnification. The left-hand view shows the entire laser structure, consisting of upper and lower mirrors, made up of alternating blue AlAs layers and green GaAs layers, surrounding a central light-emitting layer containing red indium gallium arsenide. The upper mirror requires fewer layers since a layer of gold metal at the top of this miniature soda can–shaped device acts as a final light reflector that returns even the tiny fraction of a percent of light not reflected by the upper stack. It also serves as an electrical contact to the upper half of the structure, which is doped to become a p-type semiconductor that will provide a source of holes for the central light-emitting layer. The n-type lower mirror stack provides the electrons to the central red active region. The lower electrical contact is made through the GaAs substrate at the bottom, which is also heavily doped to be n-type. The substrate contains no gold metal, and hence the fraction of a percent of light that is not reflected by the lower mirror stack escapes out the bottom of the laser, passing through the lower GaAs substrate, which the reader will remember is transparent to light at the wavelength of the InGaAs energy gap, and out the lower surface of the wafer.

In the more traditional edge-emitting geometry, the electrical function of supplying a stream of electrons and holes and the optical function of reflecting light back into the cavity are quite distinct. The former is accomplished by the proper doping of successive layers, and the latter by the cleaving of crystal facets to the sides. The two functions are even distinct from one another dimensionally, since the former occurs vertically and the latter horizontally.

In a vertical-cavity laser, the layer structure accomplishes both functions, and both happen along the same vertical direction. The complexities that result are made evident in the next level of magnification, shown in the second panel from the left in the figure on the facing page. Here we see that the alternating layers of AlAs and GaAs that make up the mirrors also contain an AlGaAs intermediate layer, colored blue-green, between the inward side of the GaAs layers and outward side of the AlAs layers. Recall from the figure on page 103 that there is a "drop" for carriers moving from higher-bandgap materials such as AlAs into lower-bandgap materials such as GaAs, so movement in that direction is relatively easy. Movement from GaAs to AlAs, on the other hand, requires "climbing" this same step, with the help of a little more voltage from the external battery. The AlGaAs layers are inserted as an intermediate step to "help" the carriers on their way, allowing the carriers to "climb up" in two small steps rather than one big one.

This first magnification level reveals another feature: the slab of InGaAs material—the active glowing portion of the overall structure—is in fact three separate layers. Strained InGaAs of this composition can be grown without defects to a critical thickness of only about 10 nanometers. But the device designer wants the highest possible light gain each pass through the structure, so the trick he or she employs is to incorporate more than one such strained InGaAs layer, for a single InGaAs layer at its maximum critical thickness would not be able alone to amplify the light traversing it enough to overcome the fraction of a percent light loss out the lower mirror stack.

These three red InGaAs layers appear more clearly at the next level of magnification (second panel from the right). It is at this level of magnification that one can also see an example of the technique of "digital alloying," mentioned briefly earlier in the chapter, in use. To make the transition less abrupt for the carriers coming in from the top and the bottom higher-bandgap blue-green AlGaAs to the central lower-bandgap green GaAs layers immediately cladding the InGaAs layer, the device designer wants to gradually grade the material from the blue-green AlGaAs to the green GaAs. He or she could ask the crystal grower to gradually cool the aluminum effusion cell so that the amount of aluminum in the layer decreases continuously, except that this procedure cannot be successfully carried out over such a short growth distance. Instead, as can be seen in the figure, this compositional grading is accomplished by cyclically chopping back and forth between the two materials, AlGaAs and GaAs (i.e., by opening and closing the aluminum shutter, something that can be performed quickly and precisely), such that there is an increasing fraction of one and a decreasing fraction of the other in each successive cycle. In such a manner, the grower is effectively grading from blue-green to green. These individual layers are so thin that the charge carriers feel the average effect of a number of them at any given time, so the digitally chopped region appears to the electrons and holes to be a smoothly graded transition from one material to the next.

The final magnification, shown at the far right, now magnified close to ten million times, begins to reveal the underlying atomic structure of the semiconductor layers. What was indicated as a green GaAs layer to the left is now seen as alternating layers of individual gallium and arsenic atoms. Furthermore, the layers of $Al_{0.5}Ga_{0.5}As$ are revealed to be alternating planes of arsenic atoms and a 50:50 mixture of gallium and aluminum atoms. At the atomic level there are no blue-green "aluminum-gallium" atoms—only blue aluminum and green gallium ones. Again, the effective range of the individual electrons and holes in the bands is much larger than the distance from one gallium

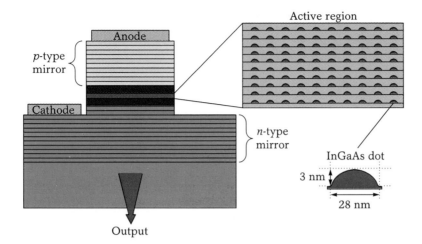

The quantum dot VCSEL, the latest version of the microlaser, has an active region that consists of an array of quantum dots instead of continuous layers. According to predictions, such laser structures should have very low threshold currents.

atom to the next aluminum one, so the overall effect on the carriers is an average aluminum gallium arsenide energy gap. In effect, the crystal grower is performing materials engineering on an atomic scale so small that the carriers he or she is attempting to control are unaware of the details!

Scientists have continued to come up with new innovations in lasers, including a laser, called the QD-VCSEL, that combines the principles of two state-of-the-art lasers in a single device. The QD-VCSEL is a vertical-cavity surface-emitting laser that makes use of quantum dots (QD) in the active layer. Researchers grow this structure by molecular beam epitaxy as in the case of the conventional VCSEL, but in place of the usual quantum well active layer, they prepare arrays of quantum dot structures. To begin the formation of the active layer, they deposit only a few monolayers of atoms on a thin buffer layer. Then, through their choice of temperature and initial substrate crystallographic orientation, they allow these atoms to self-organize to produce an array of quantum dots, in the case shown above, with a size of about 3 by 28 nanometers each. As this process is repeated, sets of QD arrays, each formed on its own continuous buffer layer, become stacked on top of each other to form the complete QD active region of the laser. Work on these laboratory prototypes is now in its infancy, and researchers are still striving to attain the very low thresholds predicted for quantum dot lasers.

The results of all this laser engineering are the smallest lasers in the world, so small that a million of them can fit on a chip the size of your smallest fingernail! Engineers are only now beginning to explore the possible uses of such devices. In the final chapter, we turn to frontiers in the ever-expanding variety of applications employing these amazing devices we call lasers.

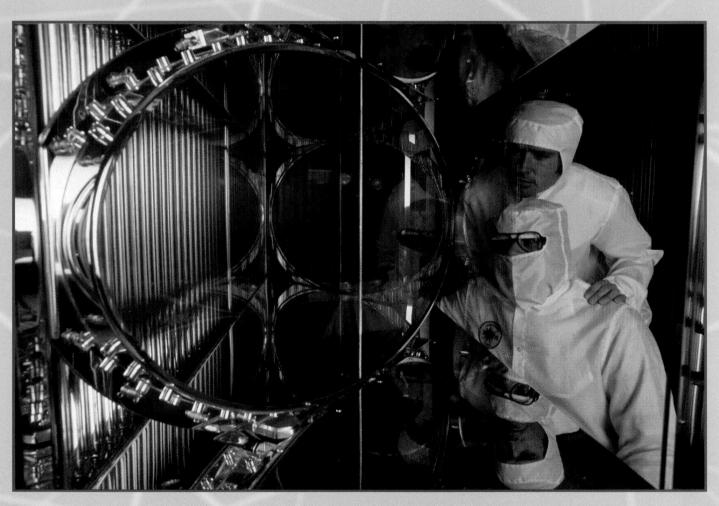

Two clean-room technicians examine a huge lens of neodymium glass. The lens forms a part of Lawrence Livermore National Laboratory's Nova laser, the world's most powerful laser.

8

Laser Frontiers

In this final chapter we turn from what can be done today with lasers to what might be done in the near future. What are the ultimate limits to how small a laser can be made? how powerful? how brief its pulses? How far can we push the lasing operation into hitherto unexplored portions of the electromagnetic spectrum? These limits are now within reach of the lasers in laboratories at the cutting edge of laser science. Their probing of these limits is opening an array of new applications that will further extend the usefulness of this highly versatile tool.

The Frontiers of Miniaturization

Since the first gas laser was set up on a laboratory bench in 1960, the smallest lasers in common use have shrunk a thousandfold in each dimension, to the size of a grain of salt. And the vertical-cavity surface-emitting lasers (VCSELs) still under development shrink the overall cavity length by another factor of 100 (as "discovered" by the ant in the photo on the facing page peering at a patchwork of small squares, each of which is an array of hundreds of individual lasers)! Yet it is unlikely that these dimensions will shrink significantly further in the future, since, with dimensions measuring a few micrometers, these typical VCSELs are already near in size to the one-micrometer wavelength of the laser light itself. Quantum mechanics tells us that light cannot be confined in spaces much smaller than its own wavelength in the material. Hence lasers significantly smaller than this crucial "dimension" of light would not be able to capture or contain the very photons needed for the lasing process.

But although this size frontier seems to be closely approaching its hard lower limit, ways to use these miniature devices have only begun to be explored. The standard thinking is to exploit miniaturization to fabricate many more individual lasers at a given time in a batch mode. These lasers would perform the same functions as semiconductor lasers in use today, but they would be cheaper and able to operate on less current.

But these microminiaturized marvels also possess properties that will allow them to perform totally new functions. One such property arises from their geometry on the underlying wafer. Their traditional edge-emitting semiconductor laser brothers, which rely on cleaved wafer edges as mirrors, need to be at the edge of a wafer. But VCSELs can in principle be interspersed at selected points across the entire surface of a "photonic" integrated circuit (as opposed to a strictly *electronic* integrated circuit), where they can serve as logical elements in more complex designs. Such designs become interesting on the frontier of optical computing, at which digital calculations are represented by the presence or absence of light rather than electric voltage or current as in electronic computers. By exchanging information from one element to the next literally at the speed of light and by avoiding electrical delays at each element, such all-optical computers could be blindingly fast. Although we are a long way from creating general purpose computing machines based on such "optical arithmetic," researchers from a number of groups throughout the world have already demonstrated the kind of optical logic that could be used to represent the logical operations such as AND, OR, and NOT that would be needed as the basis of any such optical computing arrangement.

As an example, the figure on page 174 shows the operating characteristics of a VCSEL created at Bellcore by a team led by Winston Chan,

The view in a scanning electron microscope of an ant "examining" one of a series of rectangular fields, or arrays, of vertical-cavity microlasers. One of these arrays, containing many hundreds of individual lasers, is shown in detail at the bottom of the photograph.

now at the University of Iowa, and which included one of the authors (James Harbison). This device has been specially prepared with additional *n*- and *p*-type layers within the stack so that it will latch in the "on" state when triggered by an optical pulse from another microlaser. The light from that previous laser turns on the VCSEL so that it emits light, and the VCSEL remains on even after the previous laser turns off. The "bit" of information represented by the VCSEL being in the "on" state is thus preserved in a kind of optical state memory. What makes the geometry of VCSELs so appealing for optical computing is the ability to form entire arrays of such elements. If arrays of VCSELs are stacked

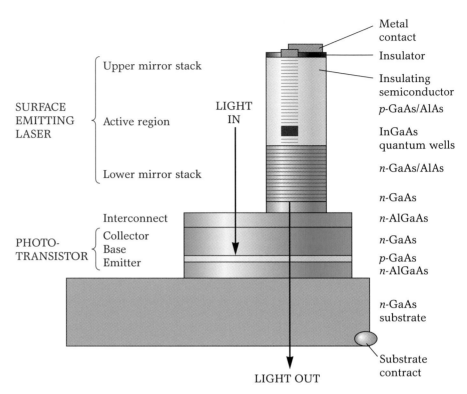

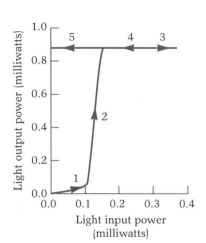

Right: A specially designed surface-emitting laser can be made to latch in the "on" state when receiving an optical input. The lower *npn* triad holds back the voltage imposed externally on the entire stack between the upper metal contact and the lower substrate contact until light from another device shines on the layers. The exposure to light creates extra carriers within those lower three layers that cause the entire structure to become conducting. The VCSEL in the stack turns on, and the lower *npn* triad is able to remain conducting. Left: Following the sequence of arrows on the plot, as the light input from the triggering device (labeled "LIGHT IN" in the diagram on the right) increases (1), the threshold point is reached at about 0.1 milliwatt, and the *npn* triad becomes conducting. The VCSEL turns on (2), firing laser light out the bottom of the overall device ("LIGHT OUT," right). As the light input continues to rise, the VCSEL turns completely on to full output (3). Thereafter, as the light input is decreased (4) and eventually turned off (5), the VCSEL remains in the "on" state.

one upon the other, VCSELs in one layer can interact by means of light pulses with those directly below them, in the next layer down. In principle each element of each huge array could be performing a function such as a logical operation or an optical memory save to the next layer below, as just described on page 173, all in parallel. For the right class of calculations, this parallelism opens the possibility of vastly increased computational speed. A two-dimensional pattern of light representing an entire image could be projected, for example, from the previous layer of processing within our optical computer to the next, and in complete parallelism, the entire "image" could be stored simultaneously.

The promise of VCSEL arrays seems particularly exciting for applications such as pattern recognition, which requires large numbers of pixels to be stored, recalled, and compared many times over for each recognition operation. The necessary calculations are slow on electronic general purpose computers, since the central processing unit of the machine must consider each pixel sequentially, but become extremely fast when allowed to occur in parallel. We are far from realizing an all-optical computer, but the potential advantages have made the development of these machines an active field of research, and an array of tiny microlasers serving as optical computer elements may be a key to opening this frontier.

VCSELs that have been fabricated into two-dimensional arrays could dramatically transform the field of visual displays, the screens and readouts we use for visualizing everything from the results of complex computer simulations to the latest TV sitcom. Most such displays are two dimensional in nature, and hence the use of some sort of two-dimensional laser array, say, as the heart of a TV or computer display device, seems like a reasonable possibility, though significantly improved optical power output levels would be required.

Prospects are just opening now for truly three-dimensional displays. Although the eye can be fooled into seeing things as three dimensional by projecting slightly different images into the left and right eyes, a long-term goal of researchers is to create displays that are themselves three dimensional, and hence as such can be examined by a group of people simultaneously from different angles. Such displays would show a true representation of all three dimensions. A research group headed by Elizabeth Downing at Stanford University has recently made a particularly interesting advance. They start with a three-dimensional cube made of transparent materials impregnated with rare earth elements that glow, or "fluoresce," when excited by intense laser light, in much the same way the phosphors on your color TV screen glow when excited by the TV tube's scanned electron beam. The key to their novel scheme is that these atoms only fluoresce when simultaneously pumped with light of two different wavelengths, coming from two separate lasers. A photon from the first laser boosts an electron within the rare-earth atom from its ground state into a particular higher-lying state. Light from the second laser, chosen to be of just the right energy, then elevates that electron to a yet higher state, from which it is then free to drop back down to the ground state, releasing the visible photon seen in the display. The two laser beams, entering the cube from below at different angles, are aimed

A prototype three-dimensional display made by Elizabeth Downing and coworkers at Stanford University and 3D Technology Laboratories. The display is built of a phosphorescent material that glows only at the intersection of a pair of laser beams of specific wavelengths. The intersecting beams act as a cursor that can be placed at arbitrary positions within the block of transparent display material. Scanning the cursor produces a three-dimensional image such as the one shown here of "Lady Liberty's" head.

This scanning electron micrograph shows a laboratory prototype of the corner portion of a square 1024 element array of vertical-cavity surface-emitting lasers (VCSELs), made at Bellcore, that are electrically addressable. All the VCSELs in a given row are connected at their bases by a stripe extending out to one of the pads at the left of the photo. In a similar way, metal stripes connecting the tops of all the VCSELs in a given column extend to the top to another series of pads. By applying an electrical voltage between a pad in a given row and another in a given column, the operator turns on the VCSEL at the intersection of that column and row. Such geometries could eventually lead to new two- and three-dimensional display technologies.

to coincide at a given, predetermined point within the cube. The result is a glowing "cursor" that can be placed anywhere within the three-dimensional cube. By scanning this laser intersection point around the cube under computer control, Downing's team has created complete three-dimensional images.

The scientists created their first pioneering display with beams from large tabletop lasers, steered throughout the three-dimensional space of the cube by means of electronically controlled mechanical mirrors. In its current manifestation, the apparatus employs much more compact semiconductor lasers, and it fits on a platform you can hold in your hand. The researchers would like to get away from the moving mirror concept altogether in favor of a more robust arrangement, and they have proposed using a pair of intersecting linear arrays of semiconductor lasers to excite the phosphors. A pair of arrays could address any point in a given plane of the display by simply firing the proper pair of lasers; no mechanical mirrors need be moved. Of course, it would require a whole stack of linear array pairs to cover the entire three-dimensional space of the display.

Imagine, though, one more step down the line in the future, a two-dimensional VCSEL array on each side of such a display. Now every three-dimensional display point could be addressed directly by firing the proper pair of lasers. Of course, there are many obstacles to the successful creation of such a device, for each VCSEL must be able to generate a great deal more power than is currently available in today's arrays and to emit light at frequencies that match the fluorescing rare earth elements in the display. Moreover, the lasers must be manufactured in electrically addressable arrays containing a minimum of tens to hundreds of thousands of laser elements; present-day laboratory prototypes, in contrast, contain far fewer devices despite their impressive appearance. These are obstacles, but not insurmountable ones, and so the effort to push on the envelope at the frontier of laser science continues.

Exploring New Parts of the Spectrum: Ultraviolet and X-ray Lasers

The visible portion of the electromagnetic spectrum to which our eyes respond as "light" is only a small slice of that spectrum. Yet all forms of electromagnetic radiation ranging in energy from radio waves all the way to gamma rays are essentially the same in their physical origin—all are manifestations of propagating waveforms of oscillating electric and magnetic fields. But however similar in essence, they differ over many orders of magnitude in the key attributes of wavelength, frequency, and energy. The laser acronym begins with "light," and yet the processes for creating an amplified beam through stimulated emission can apply in principle to the entire range of electromagnetic radiation shown in the chart on page 17. In fact, the reader will recall from Chapter 3 that the first example of lasing action was observed in a microwave device, referred to as a "maser," in the late 1950s.

In the intervening years lasers have received far more attention than masers because a wider variety of applications have been developed in the visible range. This is partly a bias resulting from our visual system. For example, sighting surveying sites or pointing at an overhead projected transparency using a laser would be useless if the wavelength of the laser was outside our eye's window of detection! But equally as important, the individual photons that make up visible light are, because of their shorter wavelength, on the order of a thousand times more energetic than those in the microwave region, unlocking powerful applications, such as laser surgery or the rapid heat treatment of metallurgical alloys, unavailable to masers. And this advantage has spurred researchers to attempt to push lasers to ever shorter wavelengths, thus creating more and more energetic photons first in the ultraviolet and now even into the X-ray region of the spectrum.

Ultraviolet light has been the object of investigation for the longer period of time. In fact, experimental prototypes of devices known as eximer lasers, created in the late 1970s and early 1980s, have now led to commercially available products. The term "eximer" is short for "excited dimer" and refers to the fact that the excited entity that gives off a photon when stimulated is not an atom, but instead a dimer molecule, a molecule consisting of two atoms bound together. Common examples of stable dimer molecules are nitrogen, N_2, and oxygen, O_2, abundant in the air around us, but in excimer lasers, to achieve the significantly higher energies of the ultraviolet portion of the spectrum, laser chemists have turned to molecules that are not stable even in their ground state.

We saw in Chapter 4 that the noble gases of the eighth column of the periodic table have their outermost electron shells completely filled, depriving them of a driving force to react to or bond with other atoms. This is the origin of their designation as the "inert" gases. But if they are excited in an electric discharge like that used to pump a typical gas laser, a remarkable change occurs. In their excited state these inert gas atoms become reactive enough to combine with similarly excited gas atoms from the halogen group in column VIIB of the periodic table, such as fluorine. Argon reacting with fluorine, for example, would form an excited ArF dimer molecule, or eximer, in a very high lying excited state. As the molecule relaxes spontaneously back to its ground state, it releases this large amount of energy in the form of an ultraviolet photon. At the same time it breaks apart into its two constituent Ar and F atoms, bringing an end to the molecule. Thus the electric discharge must not only constantly excite the individual gas atoms, it must also produce a steady stream of newly formed molecules in their excited bound state. For laser operation, of course, mirrors are added at either end of the discharge tube, and the spontaneous emission of the ultraviolet photons is replaced by stimulated emission.

One of the frontier applications of high-energy ultraviolet eximer lasers is in the field of microlithography, the photographic "printing" of patterns used to define the ultrafine features of integrated circuits. As the size of the individual transistors and the conduction paths interconnecting them has shrunk with time, these circuit features, now on the order of a fraction of a micrometer, have approached the size of the wavelength of the light used to define them by photographic exposure. Yet the trend toward miniaturization continues, so that, in order for feature sizes to continue to shrink, the need has arisen for intense

The fact that most materials absorb ultraviolet light within a very short distance forms the basis for a technique used to clean paintings. The dirt-laden surface layer of a painting absorbs the energy of the ultraviolet photons created by an eximer laser and is consequently heated to the point where it, and the dirt, is vaporized, thus removing the dirt from the painting's surface. The photograph shows areas both before (left) and after (right) cleaning.

light sources of shorter and shorter wavelength. Eximer lasers are one promising candidate for such bright light sources. KrF laser systems, now available commercially, lase at 248 nanometers, and ArF laser systems that lase at 193 nanometers are becoming available. These lasers push the limits of lithography down into the 0.2-micrometer (200-nanometer) range needed for the march to smaller and smaller features to continue. In fact, the patterning field represented by such advances is beginning to be referred to not as "microlithography," but as "nanolithography," thanks in part to the technological advances eximer lasers contribute.

Whereas visible light can penetrate some distance into many materials, most materials will have completely absorbed ultraviolet light after it has penetrated only a very short distance. This property makes eximer lasers ideal candidates for certain materials-processing techniques. Lasers may be shined onto the surface of industrial parts to harden the surface through rapid exposure to heat or even to cut through entirely in a process referred to as "laser machining." In the case of surface-hardening heat treatments, the short absorption length ensures that the energy is deposited near the surface, whereas in machining operations it ensures that the vaporized removal of material from the cut is as precise and localized as possible.

This application has been extended to the "machining" of the natural lens in the human eye to correct nature's aberrations and imperfections in the shape of the lens. An ophthalmologist is able to sculpt the lens by selectively thinning portions of it a layer at a time, using pulses of ultraviolet light from an eximer laser. These pulses, because of their short wavelength, do not pass through the visibly transparent lens, which would of course cause severe damage to the light-sensitive retina within the eye, but are instead absorbed in its surface layer. The heated surface layers are vaporized away, and the eye surgeon is able to exercise extremely fine control of the precise shape and thickness of the patient's remaining lens. The procedure permanently corrects the patient's eyesight so that he or she no longer needs glasses or contact lenses.

Later to develop have been lasers of even higher energy, in the X-ray region of the electromagnetic spectrum. Here the challenge is to find excited states of atoms that have an energy ten to a hundred times higher than even those of the excited dimers used in excimer lasers. Although X-ray lasers are not yet commercially available, in the last ten years or so researchers have been able to create working prototypes in the laboratory. The key to obtaining atoms with such a high excitation energy can be found in a discussion we had back on page 75, when we reviewed the creation of ions from atoms by removing electrons. As each electron is removed from the atom, the remaining ion becomes more positively charged. As a result, the energy required to re-move the next electron goes up significantly. This is true as well for the amount of energy needed to promote the outermost electron to a higher energy level.

The ante goes up as the ion becomes progressively more ionized. As their excited light source, X-ray lasers use atoms well down in the peri-odic table, such as selenium, yttrium, or molybdenum. These atoms are chosen since they have a large number of electrons and protons, and be-cause a great many electrons must be removed to form highly ionized atoms. The atoms need something on the order of 10 of their electrons stripped away, a state of affairs that almost never occurs in nature except in an intensely hot plasma. Furthermore, to achieve lasing the prepon-derance of the atoms need their outermost remaining electron promoted to a highly excited state, which unfortunately lasts for only a very brief time. Atoms in this state decay ten thousand times faster than our famil-iar neon atoms in the gas laser! The energy that needs to be deposited into such a system to elevate atoms to this highly excited state is orders of magnitude larger than that required in a simple gas laser, and it must

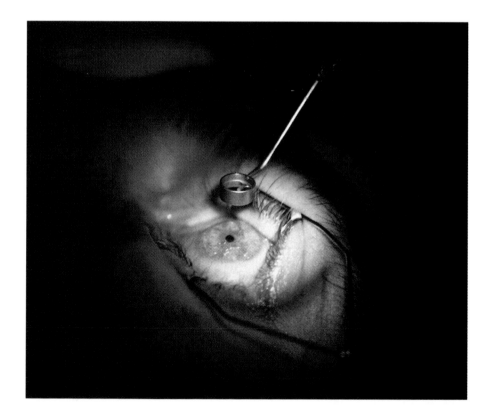

The myopic eye of a patient receives the benefit of lens reshaping carried out using an eximer laser under the eye surgeon's careful control.

This photo of one of the world's first X-ray lasers, taken in 1984 at Lawrence Livermore National Laboratory, shows a postage-stamp-size thin film target of selenium excited by the pulse from a powerful external laser along the thin line shown glowing in the photo. The excited atoms generate soft X-ray photons, at a wavelength of 20 nanometers, that are created through stimulated emission along the length of the excited region and exit through center holes in the disks at both ends of the target.

be pumped in within a time frame orders of magnitude shorter. It is no wonder such conditions took so long to achieve.

X-ray lasers have been discussed as possible defense weapons to be used as part of the Strategic Defense Initiative. They would be employed in space to fire at and destroy incoming intercontinental ballistic missiles before the missiles reenter the atmosphere. Here the proposed excitation device is a nuclear explosive, the ultimate in creating a lot of energy in a short period of time. A bundle of thin fibers made of one of the metals discussed above might be situated around the exciting nuclear device. The incredible power of the nuclear explosion would vaporize the metal, strip the necessary electrons from its atoms, and excite the remaining ions to the required elevated energy state, all in the extremely short time frame required. The X-ray photons released along the length of each wire would then stimulate more photons in a lasing process until the vastly amplified beam escapes out either end of the bundle. If the wires are long enough, the buildup of photons can occur in a single pass so that no mirrors are required. Successful operation without mirrors is essential because highly reflecting X-ray mirrors are currently unavailable, and, even if they were, they would be destroyed by the exploding device. Although prototypes have been tested in underground sites, one certainly would not want to use such a device in a laboratory or industrial application!

More down to earth is the progress that has been achieved since the mid-1980s at Lawrence Livermore National Laboratory. There the Nova laser, employed in the laser fusion application discussed at the end of this chapter, replaces the nuclear explosive device employed in the "Star Wars" design. This laser, the most powerful optical laser in the world, is capable of delivering a pulse of light energy for slightly less than a nanosecond at a power during the peak of the pulse ("peak instantaneous power") of approximately a million million watts (a terawatt)! By focusing the pulse onto a thin deposited foil of selenium along a thin line a tenth of a millimeter wide and a few centimeters long, researchers at the lab have been able to create X-ray lasing output in a manner similar to the wire-bundle nuclear powered device discussed in the previous paragraph but in a much more controlled way.

This early success led to a flurry of developments over the previous decade, and although an X-ray laser you can buy off the shelf may be a few years away, a "tabletop" model usable in a laboratory or industrial setting may be just around the corner. (The Nova laser pump occupies a length on the order of an entire football field to achieve sufficient gain in optical energy, hard to qualify as a tabletop device!) Two major advances have brought a practical X-ray laser closer to realization.

The first has been the refinement of possible X-ray mirrors, which would boost the output of such a laser while decreasing the amount of energy that has to be pumped in. Although no materials have been found that completely reflect X-rays the way a polished metallic surface completely reflects visible light, a partial X-ray reflection occurs whenever the wave encounters an interface between a material that scatters X-rays very little and a material that scatters X-rays a great deal. This phenomenon is analogous to the partial reflection of visible light off a window pane at the interface between the air and the glass. Although a single such partial X-ray reflection is not enough to make an effective X-ray mirror, we can build up the overall reflection by creating a stack of such reflectors, each carefully spaced a half wavelength apart, as we saw earlier in the figure on page 162. The concept is precisely that used for VCSEL mirrors. For X-ray mirrors, molybdenum carbide and silicon are the alternating materials typically used. The heavier molybdenum atoms scatter X-rays much more strongly than the lighter silicon ones. Again, as in the VCSEL case, the spacing from one layer to the next must be precisely half a wavelength, although in the case of X-rays, because of their shorter wavelength, this dimension is one or two orders of magnitude smaller than in the optical case, requiring exact control of layer thicknesses at atomic dimensions.

The second advance has been in the design of the device to pump energy into the X-ray laser, the special pump laser that sends a pulse of light onto the strip of selenium. What is required is an extremely high peak power in order to generate the excited vaporized metal ions and a rapid pulse to get the short-lived, higher-lying energy levels all populated at once. The strategy that has been successfully pursued toward these ends in the past few years has been to make the excitation pulses shorter and shorter. Since the energy delivered by a pulse is the product of the power and the time, a short pulse can achieve the same peak power as a pulse that is a hundred or even a thousand times longer, while its total energy is a hundred or a thousand times smaller if its pulse length is that much shorter. X-ray lasing can thus become feasible in a small laboratory lacking the extreme power of a system like the Nova laser. This push to ultrashort pulses, a frontier in and of itself, has in recent years made possible great strides toward the achievement of laboratory-size X-ray lasers.

A host of applications await the development of these X-ray lasers. Many would make use of the extremely high peak power these lasers generate for the brief period they are on, a result of the high energy of their individual photons and the extreme shortness of the pulse into which these photons are compressed. As an intense source of X-ray radiation for laboratory studies, they exceed, by a factor of one hundred million, the peak brightness level of today's most intense X-ray sources, provided by specially designed electron particle accelerators, although admittedly they provide this intensity of power for only a very brief period of time, while the accelerators emit X-rays continuously without pause.

One particularly exciting prospect is the creation of X-ray holograms. The resolution of a hologram is ultimately limited to the wavelength of light being used, on the order of a micrometer or so for visible light. X-ray holograms would open up an entirely new realm of microscopic holography and have the added advantage of being able to freeze the image in time on the order of the length of the X-ray laser pulse—less than a billionth of a second. Visionaries are already talking of capturing X-ray holograms of living cells and other complex microstructures.

Longer–Wavelength Lasers

Work is also proceeding to explore wavelengths longer than those of visible light. Most semiconductor communications lasers operate in the part of the infrared spectrum just beyond the visible portion in wavelength, known as the near infrared, near the 1.5-micrometer wavelength. Far beyond this part of the spectrum, in the microwave portion, lie the millimeter wavelengths emitted by masers, the devices that first demonstrated the lasing phenomenon, in the late 1950s. In the intervening years, there has been a significant effort to fill in the gap between the near infrared and microwave portions of the electromagnetic spectrum. One of the early gas lasers worked in a part of the infrared spectrum within this gap. This carbon dioxide, or CO_2, laser employed an excited molecule as its source. In addition to the atomic energy levels discussed in Chapter 4, a molecule also possesses a series of quantized levels related to the vibrations between the constituent atoms bonded to one another. These vibrational energy levels, in addition to an even denser array of quantized levels associated with the spinning or rotational motion of the molecule, form a rich spectrum of closely spaced levels in the infrared. Transitions between such rotational-vibrational levels are the source of the CO_2 laser's 10-micrometer-wavelength output.

Even farther out in the spectrum toward microwaves, at wavelengths in the far infrared near 100 micrometers, could be found the emissions of early lasers based entirely on transitions between molecular rotational levels. The molecules employed included hydrogen cyanide and even water vapor! But one of the most exciting recent entrants into this field is a new kind of laser called the free electron laser. Although some of its most interesting applications are in the far infrared, in principle it can work at wavelengths all the way from the microwave region up through the optical and even potentially into the near ultraviolet.

A free electron laser achieves its phenomenal tunability by using as its light source not transitions from one fixed energy level to another but the excitation of unbound electrons. Electrons that are forced to oscillate periodically back and forth in a given direction produce an alternating electric field, which in turn creates electromagnetic radiation that spreads out in the plane perpendicular to the oscillation. This is the principle behind a radio transmitter tower: the electrons are driven up and down at the radio frequency assigned to the given radio station, and their oscillation creates radio waves, a low-frequency form of electromagnetic radiation. These radio waves spread out in a plane parallel to the ground and induce a vertical oscillation of electrons at that same frequency in any vertical antenna they encounter, say the one in your automobile. That frequency is selectively tuned into by your radio receiver.

The figure on the next page illustrates the basic principle of operation of a free electron laser. A beam of electrons, accelerated to the appropriate speed, is sent down a channel between a series of magnets of alternating field. By the basic rules of Maxwell's laws, these alternating magnetic fields cause the electrons to wiggle back and forth in the direction transverse to both their horizontal overall beam direction and the vertical direction of the magnetic fields. The result is the wavy path they trace in the figure. Just like the electrons driven up and down a radio transmitter tower, these electrons radiate electromagnetic waves in a plane perpendicular to that oscillatory motion. Some of that electromagnetic radiation travels along the axis of the two mirrors; it is reflected back along this same axis and stimulates light emission from additional electrons. In this way an electromagnetic field is built up within the cavity in exactly the same way as for the simple gas laser, and the simple addition of a partially reflecting mirror at one end leads to a nicely collimated beam.

The key difference that distinguishes the electron laser from the gas laser is that the energy of the radiation depends only on the speed of the injected electrons. By causing them to traverse the cavity more slowly or quickly, the scientist can adjust the frequency of their oscillation up or down at will. Although such a device becomes ever more difficult to operate as this frequency is increased, the electron laser has run successfully, at least at lower power levels, up through the optical range. Once this type of laser leaves the research stage, a wide range of applications beckon.

At high power, beams from the electron laser could act as a ground-based defense weapon against missile attack, if they were generated at a frequency that could propagate easily in the atmosphere — say, in the

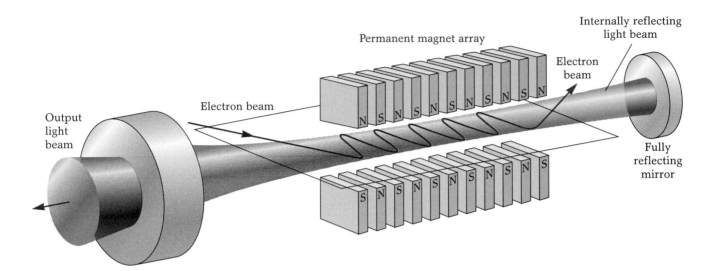

The free electron laser uses a beam of electrons as its source of the electromagnetic radiation forming its laser light. Unbound electrons are introduced into the cavity as a beam that oscillates in the alternating magnetic field generated by a series of permanent magnet pairs of alternating polarity. The frequency of oscillation in turn determines the frequency of the electromagnetic radiation that builds up in the lasing cavity between the two mirrors. The operator can continuously adjust the frequencies of oscillation, and therefore of the frequency of the resulting electromagnetic radiation, by simply varying the speed of the incoming electrons. Such lasers have been made to operate at wavelengths all the way from the far microwave to the visible regions.

near infrared at about one-micrometer wavelength. More interesting applications take advantage of the laser's ability to generate intense light in the far infrared part of the spectrum. The free electron laser facility at the University of California, Santa Barbara, shown in the photo on the opposite page, can operate in a wavelength regime of many hundreds of micrometers. Scientists at that university have come up with a number of ways to use such a tunable laser to study phenomena of interest. They have, for example, looked at how likely it is that light of a certain wavelength will induce mutations in DNA molecules. And they have studied the vertical transport of electrons through multiple stacks of alternating semiconductor materials such as the gallium arsenide and aluminum arsenide used in VCSEL mirrors. The incoming radiation induces an oscillating electric field in the multilayer stack at the frequency of the infrared radiation, which is on the order of trillions of times a second (terahertz) for the far infrared. This rapidly changing field induces an oscillation in the electrons within the stack and, if the stack is properly oriented, causes the electrons to move rapidly up and down through the multilayer interfaces. Such experiments provide a unique way of directly measuring vertical electrical conductivity in alternating stacks of materials of interest for applications such as VSCELs, and allows such measurements to be made at terahertz frequencies, faster than any conventional electronic measurement instrumentation can operate. Here, as in the case of the DNA studies, the ability to freely adjust the frequency of the radiation is an important advantage. Scientists have even proposed injecting free electron laser beams with millimeter wavelengths into a magnetically contained plasma in which scientists are attempting to induce thermonuclear fusion. The beams might supply enough extra heat to kick off the desired thermonuclear reaction.

An overall view of the free electron laser facility at the University of California, Santa Barbara, gives an awesome sense of the complexity required to successfully operate such a laser. The beam of electrons enters from the upper right corner of the photo from an accelerator behind the rear wall. It makes a pair of right angle turns, bent by magnets, to feed one of three free electron lasers that run side by side along the left side of the photo. The left beam operates in the wavelength range of from 2500 to 338 micrometers, the middle from 338 to 63 micrometers, and the right, currently under development, all the way down to 30 micrometers. Stretching along most of the straight lengths of each laser are the pairs of magnets used to undulate the electron beam. The computer monitor in the right foreground gives a sense of the large scale of the facility.

The Frontier of the Ultrashort

Another vital frontier in laser research is the generation of shorter and shorter pulses of laser light for probing and even manipulating some of nature's fastest processes, such as the vibration of atoms or the formation of chemical bonds. One way of creating short laser pulses is to use a technique referred to as "Q-switching." The "Q" of a laser cavity is a quantitative measure of how well it reflects light back and forth within itself, and hence how efficiently it will stimulate the excited atoms or molecules within it to emit photons. In a Q-switched laser, some form of device is inserted into the laser cavity to hold the Q artificially low while the population of excited atoms builds to a maximum. When the Q-switch is abruptly switched on, the entire population is triggered to lase in a very short period of time. The laser emits a short intense pulse. Under the right conditions, these pulses can be made as short as a few billionths of a second (nanoseconds). This slice of time is certainly brief, but it actually seems prolonged when you consider that the length of a one-nanosecond pulse, traveling of course at light's speed of 3×10^8 meters per second, is a third of a meter long and consists of close to a million successive wave crests!

A second tool used for achieving even shorter pulse length is a technique known as "mode locking." The name originates from the fact that, for a given laser cavity, there are a whole series of possible wavelengths that meet the requirement that as they travel the length of the cavity and back, they return back on themselves at an exact integral number of wavelengths, akin to the allowed electronic Bohr orbits we discussed in Chapter 2. There is a mode, for example, with exactly 500 peaks, one with 501 peaks, 502, 503, and so on. Suppose that we adjust the phase of all the modes so that they start out at the same point, say the top of their cycle, and let them all propagate freely throughout the cavity and back. Over the entire length of the cavity the waves will be interfering randomly with one another in both the upward and downward direction, and the average added value of all of them will be very small. This is true of every place except the original spot at which we started them. At that spot each will have traveled a full round trip, and hence they will all return at their peak value to add together constructively.

Suppose we insert a shutter next to one of the mirrors within the cavity along the axis of the laser. No light in any of the modes is allowed to build as long as it is closed. When we finally open it, however, the electric fields of all the possible allowed cavity modes are suddenly allowed to grow and peak in this previously forbidden portion of space. We quickly close the shutter again, but not before light in each of the allowed modes, all in phase and representing a strong pulse, begins traveling down the cavity length away from the shutter. That pulse travels down the tube at the speed of light, hits the mirror, and returns to the shutter after a precise period of time. To reinforce the pulse, we briefly open the shutter again just as it arrives at the shutter opening, where the peaked fields are all adding up at that point in space, then close the shutter again. The mode-locking process thus suppresses all light but that in the very short pulse. The pulse continues to circulate back and forth along the tube axis, building each time in amplitude, yet remaining extremely short. In fact, because of the way the modes all add together, the shortness of the pulse turns out to vary inversely with the number of allowed modes in the cavity, so that for long cavities, optimized for such a process, the multitude of closely spaced modes results in pulses that are exceedingly short. This is the principle of mode locking, and it has been used successfully for many years. With added enhancements, such as pulse compression elements that act in such a way as to slow the front portion of the pulse down and speed the rear edge up, pulses can be created that are as short as a few million billionths of a second (femtoseconds)!

The current record for the shortest pulse, achieved in this way, measures less than 5 femtoseconds, which corresponds to a wave only about a micrometer long. At this length the electric and magnetic fields oscillate up and down only a few times! In fact, radiation any shorter than a complete cycle is by definition no longer a wave, so the length of such pulses is now reaching an ultimate lower limit. Yet there still exist new frontiers at this limit, and these are attracting a great deal of scientific attention. One frontier is the simple manipulation of pairs of such pulses to create a tool for making measurements in the laboratory at ultra-short time resolutions.

For example, if a property such as the state of excitation of a solid can be probed optically, say by measuring the light absorption of the solid, then ultrashort pulses allow that property to be measured at a very precise point in time. In studies employing pairs of ultrashort pulses, often called "pump-probe" experiments, the first pump pulse places the solid in some transient excited state, and the probe pulse, which follows shortly thereafter, optically measures how the property of interest, say the absorption of the solid in the example just cited, has been affected by the initial pump pulse. How much of the second pulse gets through the solid in our example would yield the optical absorption of the solid, and it could measure this property at a very precise point in time after the initial excitation. Both the pump and probe pulses are formed from the same initial ultrashort pulse by splitting the beam into two parts that subsequently travel separate paths. Through the use of a movable mirror to lengthen the distance the probe pulse travels relative to the pump pulse before hitting the target solid, the time delay between the arrival of the two pulses can be precisely varied. Thus scientists can determine how the measured property continuously changes after the initial excitation by the optical pump pulse over a time frame only slightly greater than the ultrashort duration of the pump pulse itself by performing the experiment many times, each with a slightly different delay time obtained by carefully adjusting the probe pulse path length.

Such techniques have been applied to measure the ultrafast decay of excited electrons in materials such as semiconductors. On this time frame, the vibrations of atoms in molecules, which typically vibrate at a frequency of about ten million million times per second, are frozen in time! Chuck Shank and a team of scientists at the University of California at Berkeley have used such techniques to study the absorption of light by the eye's photoreceptor molecule rhodopsin; they found that on receiving the optical input representing the light the eye's photoreceptor is designed to detect, the molecule twists into a new shape on the time scale of 100 femtoseconds.

As was alluded to earlier in this chapter, the achievement of ultrashort pulse lengths could make tabletop X-ray lasers a reality in the near future. The idea is to make the pulses shorter and shorter while keeping the energy per pulse constant. In this way one can create incredibly high instantaneous peak powers, on the order of terawatts (a million million watts) for pulses of hundreds of femtoseconds or less, enough energy to initiate lasing in an X-ray laser. These high-powered lasers, creatively applied, may also bring about the appearance of new phenomena not seen at lower power. One example is being studied at the Center for Ultrafast Optical Science at the University of Michigan, where a team led by Donald Umstadter has observed that the extraordinarily high electric fields created by intense laser power can produce a plasma of charged particles out of the gas through which the beam passes. In a surprising discovery, these scientists have found that such a thread of ionized plasma is able to accelerate a beam of electrons to extremely high energies, as much as a billion electron volts after only a centimeter of acceleration. They have also made progress recently in forming much longer plasma channels, making use of self-organizing interactions between the laser light and the plasma to keep the light tightly focused. There is hope that such techniques will lead to tabletop accelerators comparable in energy to the most powerful conventional particle accelerators covering acres of land!

One of the age-old dreams of laser chemistry has been to create a means of breaking a particular bond in a molecule, thus allowing chemical reactions to be controlled with unheard-of precision. Much early work in lasers centered on trying to tune a laser to a bond's particular frequency of vibration, in the hope that applying the tuned laser would cause the bond to oscillate wildly and break apart. Despite a couple of decades' work along these lines, such hopes have gone essentially unfulfilled. Although such a scheme can inject a lot of energy into the molecule by exciting a given bond, in general this excess energy is unfortunately rapidly distributed to all the other bonds in the molecule and thus transformed into heat before it has time to break the particular bond for which it was intended.

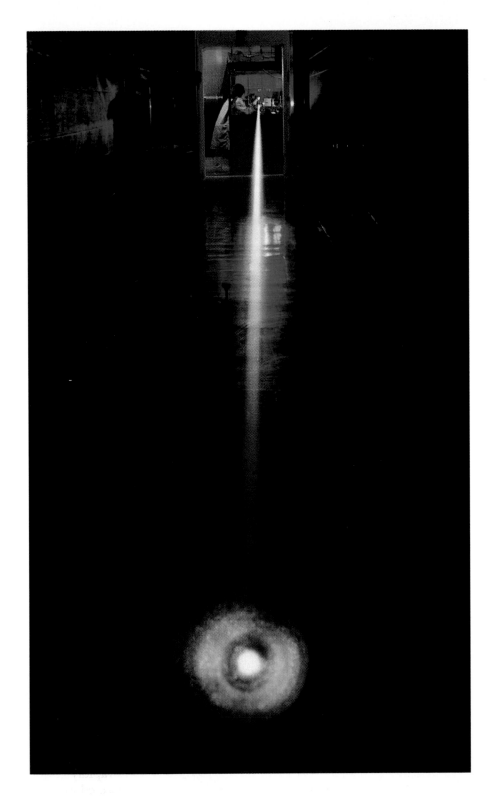

A self-guided laser pulse of ultrashort duration at the University of Michigan creates a 40-meter plasma channel. Such a stream of plasma may someday be used to optically accelerate electrons.

Advances at the ultrafast pulse frontier now appear to be breaking this impasse. Researchers are developing sophisticated control not just of the pulse length, but also of the shape of the pulse over time. By adjusting the phase, frequency, and amplitude of the pulses, the laser chemist can control the complete interaction of the pulse with the molecule. When the molecule is exposed to the light pulse, different bonds in the molecule respond to different portions of the complex pulse and hence respond at different times. Think of the process as grabbing onto the molecule and shaking it at multiple points at carefully controlled intervals. The interactions resulting from these series of excitations can, in principle, be tailored to add up together in one place, causing, for example, a given atom to be ejected from the molecule. Such advances are beginning to happen due to the confluence of the ability to control the shape of ultrashort pulses and the increased theoretical understanding of the complex quantum mechanical effects of such excitations on the target molecules.

Atomic Trapping and Cooling

One of the more novel frontiers in laser research is the use of lasers to levitate, trap, and manipulate objects ranging from tiny plastic balls to individual atoms. This line of research dates back almost to the very earliest days of lasers. As early as the late 1960s and early 1970s, Art Ashkin at Bell Laboratories levitated and manipulated tiny objects with lasers. Because individual photons each possess a small amount of momentum, an object such as a small, light sphere that absorbs a photon or has a photon reflected off it will in turn experience a change in its momentum, just like balls colliding on a billiard table. Ashkin showed that, by imparting momentum of the proper magnitude and direction, a focused laser beam could trap a small plastic sphere, about a micrometer in diameter, floating in a suspension in water, and under the right circumstances could pin the sphere against a window on the wall opposite to that in which the laser beam entered the water chamber. He then came up with a scheme for shining a pair of lasers simultaneously in from either side of the chamber to trap balls in the chamber's center.

A number of research groups extended this work in the 1980s to attempt to manipulate and trap individual atoms. They chose the frequency of the laser beam to match the energy of a particular energy-level change within the atom to enhance the photon's absorption. Furthermore, by tuning the laser slightly off this frequency, they gained the ability to interact only with moving molecules, whose resonant frequencies are shifted by means of the same familiar Doppler shift that affects the pitch of the sound you hear from a passing vehicle. The light could be made to slow atoms moving toward it but then stop its effective push on them as they became stationary. In such a manner scientists were able to bring beams of incoming gas atoms to a virtual standstill, effectively cooling them to within a degree of absolute zero (1 degree Kelvin). The atoms could then be trapped using a variety of techniques, including intersecting beams such as those employed earlier by Ashkin. More sophisticated magnetic traps allowed the warmer of the trapped atoms to escape, leaving behind cooler atoms, in the same way that your forehead is cooled as the sweat on your brow evaporates, leaving behind the cooler water molecules. This technique achieved temperatures on the order of a millionth of a degree above absolute zero (a microkelvin). The photo on this page shows the apparatus employed in Steven Chu's laboratory at Stanford University, where he and his group are able to trap atoms and even maneuver them into a special upper portion of the chamber in which they can be subject to further precise measurements.

This apparatus at Stanford University uses a combination of laser beams, shown shining into the vacuum chamber, and magnetic fields to trap atoms for study at extremely cooled temperatures, on the order of a millionth of a degree above absolute zero. The stream of blue dots of light entering the chamber from the right side of the photo are pulses of ultraviolet radiation used to probe the trapped ultracold atoms.

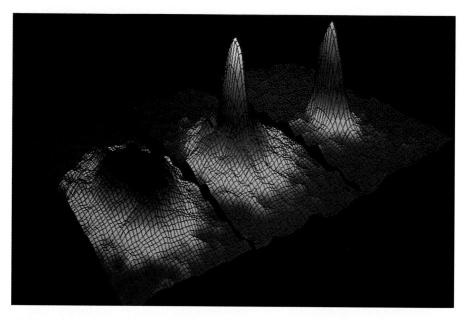

This series of three successive measurements of the velocity distributions in a cloud of laser-trapped and magnetically cooled rubidium atoms provides conclusive evidence for the appearance of a Bose-Einstein condensed state of matter. This exotic form of matter has been observed only below the incredibly cold temperature of 170 billionths of a degree above absolute zero. Each of the three plots is obtained by taking a snapshot of the particle positions a fixed time after the forces holding the atoms in the trap are removed. The slower particles remain near the center, while faster ones reach outward to the edges of the rectangle. The height of the displayed curves indicate the number of particles found at each of these velocities. These heights are made more apparent by the use of a series of colors reflecting the height, ranging from red to orange to yellow to green to blue to white. The left plot shows the cooled atoms just before Bose-Einstein condensation, the middle plot shows the atoms just after the condensate appears as a peak of particles at near zero velocity in the lowest quantum state, and the right plot, taken shortly thereafter, shows nearly pure condensate, left behind as the noncondensed material with higher velocities evaporated away.

One of the fascinating predictions of quantum mechanics is that a certain class of particles known as bosons will in principle, when cooled to a low enough temperature, condense into a state in which every particle is in the identical ground state. Bosons are one of two general categories of particles described by quantum mechanics, the other being called "fermions," a category that includes familiar particles such as electrons, protons, and neutrons. Fermions are particles no two of which can simultaneously occupy the same quantum state. It is this exclusivity property that causes electrons to fill the energy states of an atom from the bottom up.

Bosons, on the other hand, have no such exclusivity restriction. Just because one boson is in a particular quantum state does not preclude others from joining it. Referring back to the amphitheater analogy of Chapter 2, the behavior of bosons could be compared to the entire audience all choosing to sit together in the closest seat of our amphitheater! A familiar example of a boson is a photon. Just because one photon is in a state reflecting back and forth between the two mirrors within a laser cavity does not preclude others from joining it. If this were not true, lasers would never work.

That bosons could all condense into this lowest-lying state, all into the closest seat in our analogy, was independently predicted by both Albert Einstein and the Indian physicist Satyendra Nath Bose in 1924, and the process is commonly referred to as Bose-Einstein condensation. In the seventy or so years since that prediction, only a few macroscopic systems have been discovered that represent such condensation. Superconductivity in certain metals at low temperatures is such a Bose-Einstein condensate, and its interesting properties have captured the imaginations of both physicists and the general public alike. In principle, certain atoms and molecules, acting as bosons, should condense in such a way at a very low temperature if cooled while still suspended in a noninteracting gaslike phase.

The prospect of observing such condensation drove a number of research groups in the early 1990s to perfect the techniques for trapping and cooling atoms, which we have just been discussing. In the midst of the "heated" competition among research groups, in July 1995 a team announced to the world that they had successfully achieved Bose-Einstein condensation. This team included Eric Cornell of the National Institute of Standards and Technology, Carl Wieman of the University of Colorado, and their colleagues Mike Anderson, Jason Ensher, and Mike Matthews at the Joint Institute for Laboratory Astrophysics, all in Boulder, Colorado. They had achieved condensation in a collection of rubidium atoms cooled below 170 nanokelvins. The figure on the opposite page shows a velocity distribution of the particles in the trap; a remarkably abrupt peak at zero velocity (the center of each rectangular plot), which corresponds to atoms in the condensate, is seen to grow in the time sequence of the three successive measurements shown in the three plots from left to right. Since then other groups have achieved similar results using other collections of atoms, and a wide array of analytical tools are being assembled to analyze and measure the properties of this new state of matter.

Ultra-High-Power Lasers

We end our account by taking a look at the final frontier, the achievement of ultra-high power. It is this frontier that has most captured the public's imagination, as high-powered beams of energy have appeared everywhere from movies like *Star Wars* to the concept of the Strategic Defense Initiative that glimmered so brightly in the plans of President Reagan during the 1980s. These beams are one science-fiction fantasy that is already becoming a reality.

One of the most intriguing uses of the awesome power of a high-energy laser is being explored at Lawrence Livermore National Laboratory, in a project begun many years ago. The project members aim to create enough power, pressure, and sheer energy in a small pellet of nuclear fuel to achieve the conditions in the center of the sun and initiate thermonuclear fusion, the reaction in which two hydrogen nuclei fuse together and provide the sun with an almost limitless source of energy. The effort has made steady progress over the years, and similar projects have already been initiated in other labs. Even now the U.S. Congress is discussing appropriations to build a facility for laser fusion called the National Ignition Facility. To be completed around 2002, it would include 192 laser beams in a building the size of a football stadium, each focused on a single central point containing hydrogen suitable for fusion. Whereas laser fusion demonstrations to date have in fact induced fusion to occur, the energy recovered from the fusion reaction has been less than that needed to fire the lasers required to induce it. The National Ignition Facility could be the first to cross the breakeven point. Its predecessor exists already today at the Lawrence Livermore National Laboratory, and we end our tour of laser frontiers by watching in awe as a single burst of the world's most powerful laser light is fired for a brief nanosecond on a solitary hydrogen pellet.

At one end of a laboratory nearly the length of football field stands a spherical stainless-steel chamber the size of a man, at the center of which sits a hollow pellet of glass only a millimeter in diameter. Within this tiny container are small amounts of deuterium and tritium, rare heavy forms of hydrogen that serve as the fuel for the thermonuclear reaction at the center of the sun. Until forced together by pressures and temperatures akin to those found at the center of the massive sun, they wait quietly, not reacting.

At the other end of the long room sits an ultra-high-power neodymium glass laser being charged up to fire. Embedded within the glass rod at the center of the laser are isolated neodymium atoms that have been pumped to an excited state by banks of flash lamps. When the light energy within the glass reaches a threshold amount, the device lases and a pulse of 1.06 micrometer infrared light is released out one end. The pulse is first sent through a complex maze of optics that evenly splits it into ten concurrent pulses, each directed down one of ten parallel optical raceways shown on the facing page. As the pulses travel the lengthy distance of the laboratory, each passes through sections of material

The world's largest laser. This ten-beam Nova laser at Lawrence Livermore National Laboratory is roughly the length of an entire football field and is capable of delivering a pulse with a total peak power of 10,000,000,000,000 watts! When focused on a small pellet of heavy hydrogen, it is able to induce nuclear fusion like that which occurs at the center of the sun.

pumped to an excited state; through further stimulated emission during passage through each section, the pulses grow in power. In addition, each pulse is transmitted through a pair of carefully oriented perfect crystals that triple the frequency of each light particle, converting the pulses into high-energy ultraviolet light. At the end of these parallel amplification paths, a network of mirrors cleverly redirects the separate pulses to converge upon the center of the reaction sphere, where the tiny glass capsule of hydrogen sits. For a few billionths of a second, all ten pulses, each at a power level of slightly more than a trillion watts, enter a small cylindrical container surrounding the pellet, whose interior walls are made of the heavy element uranium. The interaction of the UV light with the uranium atoms creates an incredibly intense bath of X-ray radiation that surrounds and impinges upon the surface of the tiny pellet containing the gas.

The outermost layer of the pellet vaporizes into a plasma almost instantaneously, streaming outward at a speed in excess of a million miles per hour. Propelled inward with an equal but opposite force, the shell collapses on itself, shrinking the size of the pellet to a fiftieth of its original diameter. In the process, the pressure on the hydrogen rises above a billion atmospheres, while its temperature soars to many tens of millions of degrees Celsius, near the temperature at the center of the sun. At the incredibly high density and temperature attained in that fleeting instant, the nuclei of adjacent hydrogen atoms are driven to within a few trillionths of a millimeter of one another, so that they are squeezed closer together than a hundred thousandth the size of an individual atom. At this short range, the forces that hold the protons and neutrons of the nucleus together become stronger than the enormous electrostatic forces that have hitherto forced the positive nuclei of the two separate hydrogen atoms apart. The two nuclei collapse together and "fuse" to form a helium nucleus, and enormous amounts of energy are released. The energy produced by this means may someday dwarf the initial input of laser light energy in the reactor, paving the way for an era of almost limitless energy, fueled by heavy hydrogen obtained from abundant supplies of ocean water.

We have seen how the basic quantum mechanical workings of atoms, when harnessed into the relatively simple device we call a laser, make possible a host of applications ranging from the everyday to the truly fantastic. The laser is a perfect example of how the results of a deep scientific quest to understand nature more fully, resulting from centuries of grappling with fundamental questions of "why it all works," can be applied to create a tool we could scarcely live without in our modern technological society. And the path to further understanding, with its promise of as yet undreamed of applications, waits quietly, invitingly, for the tread of human understanding.

Further Readings

Amorphous Solids

Bernal, J. D.: "The Geometry of the Structure of Liquids," in Thomas J. Hughel (ed.), *Liquids: Structure, Properties, Solid Interactions,* Elsevier, Amsterdam, 1965, pp. 25–50.

A reasonably readable technical review of Bernal's pioneering work on modeling amorphous liquids.

Bernal, J. D., and J. Mason: "Coordination of Randomly Packed Spheres," *Nature,* vol. 188, 10 December 1960, pp. 910–11.

One of the original reports by Bernal's group, technical, but worth looking at for those interested in the subject and in particular its history.

Zallen, Richard: *The Physics of Amorphous Solids,* John Wiley and Sons, New York, 1983.

A good survey of work in the field of noncrystalline solids.

Free Electron Lasers

Freund, Henry P., and Robert K. Parker: "Free Electron Lasers," *Scientific American,* vol. 260, April 1989, pp. 84–89.

A thorough treatment of this novel kind of laser at the Scientific American level.

General Background

Masterson, William L., and Cecile N. Hurley: *Chemistry: Principles and Reactions,* 2d ed., Harcourt Brace, Orlando, Florida, 1993.

A complete, current, college-level chemistry text, covering fundamentals such as bonding, electronic orbitals, and emission spectra.

Tipler, Paul A.: *Physics for Scientists and Engineers,* 3d ed., Worth Publishers, New York, 1991.

A complete, current, college-level physics text with good treatments of electromagnetic waves.

History of Semiconductors

Braun, Ernest: *Revolution in Miniature,* Cambridge University Press, 1978.

The first few chapters give a thorough historical treatment of scientific work on semiconductors, from Faraday's early pioneering work in the 1830s through the invention of the transistor.

Editors of *Electronics:* "An Age of Innovation: The World of Electronics 1930–2000," McGraw-Hill, New York, pp. 66–70.

Contains a good treatment of the early work on semiconductors, including the discovery of the pn junction and the transistor.

History of the Laser

Bromberg, Joan Lisa: "The Birth of the Laser," *Physics Today,* October 1988, pp. 26–33.

A scholarly but very readable in-depth history of the early years surrounding the invention of the laser.

Casey, Jr., H. C., and M. B. Panish: *Heterostructure Lasers,* Parts A and B, Academic Press, 1978.

For the technical specialist, it includes historical remarks and references to the first semiconductor lasers and the first heterostructure lasers ever made.

Einstein, A.: *Physik. Z.,* vol. 18, 1917, p.121, in German.

The original article in which Einstein discussed the concepts of stimulated and spontaneous emission of light.

Gordon, James P.: "The Maser," *Scientific American,* vol. 199, December 1958, pp. 42–50.

An early article on the maser by one of the group who originally invented it.

Schawlow, Arthur L.: "Optical Masers," *Scientific American,* vol. 204, June 1961, pp. 52–61.

An early article for the layman on what later came to be known as the laser, by one of its pioneers.

Schawlow, A. L., and C. H. Townes: "Infrared and Optical Masers," *Physical Review,* vol. 112, 15 December 1958, pp. 1940–1949.

A highly technical paper, filled with equations, but noteworthy because it garnered the authors a Nobel Prize for its correct proposal of how to make the first laser.

Holography

Caulfield, H. John: "The Wonder of Holography," *National Geographic,* vol. 165, March 1984, pp. 364–377.

A richly illustrated account covering both how holograms work and the myriad fascinating applications that make use of them.

Leith, Emmett N., and Juris Upatnicks: "Photography by Laser," *Scientific American,* vol. 212, June 1965, pp. 24–35.

An early article on holography by the two pioneers who first used lasers to create these appealing images.

Laser Distance Measurement to the Moon

Faller, James E., and E. Joseph Wampler: "The Lunar Laser Reflector," *Scientific American,* vol. 222, March 1970, pp. 38–49.

A detailed description of the original laser ranging experiment by two of the scientists involved.

Faller, James, Irvin Winer, Walter Carrion, Thomas S. Johnson, Paul Spadin, Lloyd Robinson, E. Joseph Wampler, and Donald Wieber: "Laser Beam Directed at the Lunar Retro-Reflector Array: Observations of the First Returns," *Science,* vol. 166, 3 October 1969, pp. 99–102.

The original technical paper announcing the first results of the laser ranging experiment shortly after the Apollo 11 moon mission.

Laser Fusion

Craxton, R. Stephen, Robert L. McCrory, and John Soures: "Progress in Laser Fusion," *Scientific American,* vol. 255, August 1986, pp. 68–78.

A survey of the early progress in laser fusion, covering most of the key obstacles.

Peterson, Ivars: "Sparking Fusion: A Step Toward Laser-Initiated Nuclear Fusion Reactions," *Science News,* vol. 150, 19 October 1996, pp. 254–255.

A recent update both on progress in the field and plans for the proposed National Ignition Facility.

Laser Gyroscopes

Anderson, Dana Z.: "Optical Gyroscopes," *Scientific American,* vol. 254, April 1986, pp. 94–99.

The physics behind ring laser and optical fiber gyroscopes.

Leondes, Cornelius T.: "Inertial Navigation for Aircraft," *Scientific American,* vol. 222, March 1970, pp. 80–86.

A description of how mechanical gyroscopes work to guide airplanes, written at the time that such devices were first being installed on commercial aircraft.

MacKenzie, Donald: "From the Luminiferous Ether to the Boeing 757: A History of the Laser Gyroscopes," *Technology and Culture,* vol. 34, July 1993, pp. 475–534.

A thorough history of the development of the laser gyroscope, covering not only the technical aspects but also some of the "politics" involved in its introduction.

Laser Surgery

Berns, Michael W.: "Laser Surgery," *Scientific American,* vol. 264, June 1991, pp. 84–90.

Good coverage of the varied uses of lasers in the medical world.

Laser Trapping of Particles

Anderson, M. H., J. R. Ensher, M. R. Matthews, C. E. Wieman, and E. A. Cornell: "Observation of Bose-Einstein Condensation in a Dilute Atomic Vapor," *Science,* vol. 269, 14 July 1995, pp. 198–201.

The original technical paper that reported on the breakthrough creation of a Bose-Einstein condensate using laser trapping coupled with magnetic cooling.

Ashkin, Arthur: "The Pressure of Laser Light," *Scientific American,* vol. 226, February 1972, pp. 63–71.

A fascinating account by a pioneer in the field of laser trapping, describing his early experiments trapping small spheres.

Chu, Steven: "Laser Trapping of Neutral Particles," *Scientific American,* vol. 266, February 1992, pp. 71–76.

A survey of the most recent techniques for trapping atoms using lasers.

Phillips, William D., and Harold J. Metcalf: "Cooling and Trapping Atoms," *Scientific American,* vol. 256, March 1987, pp. 50–56.

An article on early work using lasers to trap and cool individual atoms.

Wu, C.: "Physics 'Holy Grail' Finally Captured," *Science News,* vol. 148, 15 July 1995, p. 36.

A short report for the layman of the breakthrough covered in the Science article by Anderson et al. cited above.

Lasers in General

Fasol, Gerhard: "Fast, Cheap and Very Bright," *Science,* vol. 275, 14 February 1997, p. 941.

A brief journal article for readers with a general science background, focused on the properties of the latest quantum dot vertical-cavity surface-emitting lasers (QD-VCSELs).

Silfvast, William T.: *Laser Fundamentals,* Cambridge University Press, Cambridge, 1996.

An in-depth treatment of virtually every kind of laser and the physics behind each, written at a more technical level than the current book.

Smith, William V., and Peter P. Sorokin: *The Laser,* McGraw Hill, 1966.

This book, intended for the laser specialist, includes technical descriptions of early lasers, such as ruby and helium-neon, along with some historical detail.

Light

Baierlein, Ralph: *Newton to Einstein: the Trail of Light,* Cambridge University Press, 1992.

A historical survey of our understanding of the properties of light. It is mostly descriptive and conceptual, with only a little algebra as necessary.

McComb, Gordon: *Lasers, Ray Guns, and Light Cannons,* McGraw-Hill, 1997.

For those interested in doing more than reading about lasers. Includes discussion of both basic physics and project applications, as the title suggests. There are 88 projects in all, from setting up a helium-neon laser to experiments with optical interference effects, to holography experiments, and more.

Perkowitz, Sidney: *Empire of Light,* Henry Holt & Co., 1996.

A nonmathematical description of the properties of light for the general reader, with application to art.

Stong, C. L.: "Amateur scientist: Build a Gas Laser in the Home," *Scientific American,* vol. 211, September 1964, p. 227

James Harbison had to include this one. It's the one referred to in the Preface that got him going on lasers and science in general back in high school.

Materials Science

Corcoran, Elizabeth: "Diminishing Dimensions," *Scientific American,* vol. 263, November 1990, p. 122.

A journal article written for the general reader interested in science issues, discussing how materials scientists are making smaller and smaller features for quantum devices. It includes lucid descriptions of microlasers and of the ideas behind quantum wires and quantum dots.

Mahajan, S., and L. C. Kimerling (eds.): *Concise Encyclopedia of Semiconducting Materials and Related Technologies,* Pergamon Press, Oxford, 1992.

A technical but complete treatment of a number of aspects of semiconductor technologies. Of particular relevance to readers are the articles on "Liquid Phase Epitaxy" by S. Mahajan on pages 275–277 and the one on "Molecular Beam Epitaxy" by J. P. Harbison on pages 306–311.

Optical Computing

Abu-Mostafa, Yaser S., and Demetri Psaltis: "Optical Neural Computers," *Scientific American,* vol. 256, March 1987, pp. 88–95.

A survey of some of the ways optics, and lasers in particular, may play a role in future computers.

Quantum Mechanics

Darrow, Karl K.: "Davisson and Germer," *Scientific American,* vol. 178, May 1948, pp. 50–53.

A description for the layman of these two pioneers' experiments on electron diffraction told by a friend and colleague. As the editor notes, the article is "not only a description of how they made their discovery but also an essay in how the physicist approaches a problem of his exacting science."

d'Abro, A.: *The Rise of the New Physics,* Dover Publications, Inc. 1951.

Another readable treatment of twentieth-century physics, including a discussion of quantum mechanics.

March, Robert H.: *Physics for Poets,* McGraw-Hill, 1970, with later editions.

A physics textbook for use in liberal arts courses and an excellent book for the general reader. It includes useful discussion of the fundamentals of quantum mechanics in nonmathematical form, just right for poets.

McQuarrie, Donald: *Quantum Chemistry,* University Science Books, Sausalito, California, 1983.

A more rigorous mathematical treatment of electronic orbitals in atoms.

Pauling, Linus, and E. Bright Wilson, Jr.: *Introduction to Quantum Mechanics — With applications to Chemistry,* Dover Publications, Inc., 1935 original edition.

A classic college-level textbook on quantum mechanics, including an excellent description of the electronic orbitals of the hydrogen atom. This is for the serious reader. Although it does not include the detailed mathematical derivations of other texts, it makes full use of equations and mathematical notations.

Trefil, James: *From Atoms to Quarks: An Introduction to the Strange World of Particle Physics,* Anchor Books, Doubleday, 1994.

The book, intended for the general reader with an interest in science, includes an excellent chapter summarizing the concepts of atomic theory. It is written in Trefil's lucid style without use of mathematics.

Zukav, Gary: *The Dancing Wu Li Masters: An Overview of the New Physics,* Bantam Books, 1980.

A readable treatment of twentieth-century physics for the general reader interested in quantum mechanics and other modern science topics.

Three-Dimensional Displays

Downing, Elizabeth, Lambertus Hesselink, John Ralston, and Roger Mcfarlane, "A Three-Color, Solid-State, Three-Dimensional Display," *Science,* vol. 273, 30 August 1996, pp. 1185–1189.

A technical paper by Downing and her group that gives the details of the operation of her new kind of display.

James, Glanz: "Three Dimensional Images Are Conjured in a Crystal Cube," *Science,* vol. 273, 30 August 1996, p. 1172.

A summary of Downing's three-dimensional display.

Wu, Corinna: "Sculptures of Light: A New Three-Dimensional Display Turns on the Imagination," *Science News,* vol. 150, 26 October 1996, pp. 270–271.

A science reporter's description of the new three-dimensional displays described in Chapter 8.

Ultrashort Laser Pulses

Brumer, Paul, and Moshe Shapiro: "Laser Control of Chemical Reactions," *Scientific American,* vol. 272, March 1995, pp. 56–63.

A more in-depth treatment of how quantum mechanics theory and increasing control of ultrashort laser pulses is bringing progress in this field, which many chemists consider to be the "Holy Grail."

Peterson, Ivars: "Laser reaction control in hot sodium vapor," *Science News,* vol. 149, 30 March 1996, p. 197.

An interpretation for the layman of recent breakthroughs in the use of ultrashort pulses in controlling chemical reactions.

Peterson, Ivars: "Surfing a Laser Wave: Toward a Tabletop Particle Accelerator," *Science News,* vol. 149, 10 February 1996, p. 95.

A treatment for the layman describing the latest work on creating channels in air in which to accelerate particles using ultrashort laser pulses.

Vertical-Cavity Surface-Emitting Lasers

Evans, Gary A., and Jacob M. Hammer: *Surface Emitting Semiconductor Lasers and Arrays,* Academic Press, Boston, 1993.

A more detailed technical treatment of the early work on surface-emitting lasers.

Jewell, J. L., J. P. Harbison, and A. Scherer: "Microlasers," *Scientific American,* vol. 265, November 1991, pp. 86–94.

A fuller description of the author's work on the first all-semiconductor VCSELs and how they operate.

X-Ray Lasers

Hellemans, Alexander: "With Mirrors and Finesse, Labs Domesticate the X-Ray Laser," *Science,* vol. 273, 5 July 1996, pp. 32–33.

A recent survey of the latest developments in X-ray lasers.

Matthews, Dennis L., and Mordecai D. Rosen: "Soft-X-Ray Lasers," *Scientific American,* vol. 259, December 1988, pp. 86–91.

A treatment for the layman of the first experiments employing the ultrahigh-power lasers at Lawrence Livermore for pumping the first X-ray lasers.

Patel, Kumar N., and Nicolaas Bloembergen: "Strategic Defense and Directed-Energy Weapons," *Scientific American,* vol. 257, September 1987, pp. 39–45.

A summary of the findings of the American Physical Society's commissioned report on the "Science and Technology of Directed Energy Weapons." It contains sections describing the feasibility of high-power lasers, including X-ray and free electron lasers, as defensive weapons in a Strategic Defense Initiative scenario.

Sources of Illustrations

Facing page 1: Georg Fischer/Bilderberg *Page 2:* Phillippe Plailly/Eurelios/Science Photo Library/Photo Researchers *Page 5:* Adapted from James E. Faller and E. Joseph Wampler, "The Lunar Laser Reflector," *Scientific American,* vol. 222 (no. 3), March 1970, page 40. © 1970 by Scientific American, Inc. All rights reserved. *Page 6:* NASA *Page 9:* The Lighthouse by Dudley Witney *Page 10:* Roger Ressmeyer/Corbis *Page 12:* Siegfried Layda/Tony Stone Images *Page 14:* Berenice Abbott/Commerce Graphics, Ltd, Inc. *Page 18:* Erich Schrempp/Photo Researchers *Page 23:* Photo courtesy of Chris Palmstrøm, University of Minnesota, based on his research at Bellcore *Page 25:* From G. Gerhold et. al., *American Journal of Physics,* vol. 40 (1972), page 998 *Page 35:* Will and Deni Mcintyre/Photo Researchers *Page 39:* Created by John Talbot (jtalbot@achilles.net). © Element spectra courtesy of the Astronomical Data Center and the National Space Science Data Center through the World Data Center A for Rockets and Satellites. *Page 40:* Ken Karp *Page 42:* Bellcore *Page 47:* Spectra Physics/Laser Plane *Page 48:* Rofin-Sinar, Inc. *Page 53:* Photo courtesy of Hughes Research Laboratories *Page 54:* Property of AT&T Archives. Reprinted with permission of AT&T. *Page 59:* Adapted from Dana Z. Anderson, "Optical Gyroscope," *Scientific American,* vol. 254 (no. 4), April 1986, page 94. © 1986 by Scientific American, Inc. All rights reserved. *Page 60:* Photo supplied by Litton Guidance and Control Systems *Page 62:* Adapted from Emmett N. Leith and Juris Upatnicks, "Photography by Laser," *Scientific American,* June 1965, page 28. © 1965 by Scientific American, Inc. All rights reserved. *Page 63:* Adapted from Keith S. Pennington, "Advances in Holography," *Scientific American,* February 1968, page 43. © 1968 by Scientific American, Inc. All rights reserved. *Page 64:* Philippe Plailly/SPL/Photo Researchers *Page 66:* INTEL CORPORATION, Company design. Diagram for Intel386 (TM) microprocessor chip. (1985) Computer-generated plot on paper, 84 × 79" (213 × 200.5 cm). The Museum of

Modern Art, New York. Gift of the Manufacturer. Photograph © 1997 The Museum of Modern Art, New York. *Page 86:* 1994 Jim Foster/The Stock Market *Page 92:* Phil Jude/SPL/Photo Researchers *Page 105:* Property of AT&T Archives. Reprinted with permission of AT&T. *Page 109:* Property of AT&T Archives. Reprinted with permission of AT&T. *Page 110:* British Telecom *Page 112:* Computer-generated graphic courtesy of Chris Palmstrøm, University of Minnesota *Page 114:* © British Museum *Page 115:* Science Museum/Science and Society Picture Library, London *Page 118:* Astrid and Hans Frieder/SPL/Photo Researchers *Page 120:* Malcolm Fielding, The BOC Group PLC/SPL/Photo Researchers *Page 121:* Sumitomo Electric, USA, Inc. *Page 138:* VG Semicon *Page 145:* Adapted from S. Fafard, K. Hinzer, S. Raymond, M. Dion, J. McCaffrey, and S. Charbonneau, "Red-Emitting Semiconductor Quantum Dot Lasers," *Science,* vol. 274, 22 November 1996, page 1351. *Page 149:* VG Semicon *Page 150:* Bellcore *Page 154:* Bellcore *Page 161:* From Jack L. Jewell, James P. Harbison, and Axel Scherer, "Microlasers," *Scientific American,* vol. 265 (no. 5), November 1991. © 1991 by Scientific American, Inc. All rights reserved. *Page 164:* Photo courtesy of Axel Scherer, Caltech, based on his research at Bellcore *Page 165:* Photo courtesy of Axel Scherer, Caltech, based on his research at Bellcore *Page 166:* From Jack L. Jewell, James P. Harbison, and Axel Scherer, "Microlasers", *Scientific American,* vol. 265 (no. 5), November 1991, p. 92. © 1991 by Scientific American, Inc. All rights reserved. *Page 169:* Adapted from G. Fasol, "Fast, Cheap, and Very Bright," *Science,* vol. 275, 14 February 1997, pages 941–942. Original source: H. Saito, K. Nishi, I. Ogura, S. Sugou, Y. Sagimoto, *Applied Physics Letters,* vol. 69, 1996, page 3140. *Page 170:* Roger Ressmeyer/Corbis *Page 173:* Photo courtesy of Axel Scherer, Caltech, based on his research at Bellcore *Page 174:* Figure by Winston Chan, University of Iowa, based on his work at Bellcore

Page 175: Photo courtesy of Elizabeth Downing, 3D Technology Laboratories, Inc. *Page 176:* Photo courtesy of Axel Scherer, Caltech, based on his research at Bellcore *Page 179:* Photos courtesy of Lambdaphysik and F.O.R.I.H. Heraklion, Kreta *Page 181:* Penny Tweedie/Tony Stone Images *Page 182:* Lawrence Livermore National Laboratory *Page 186:* Adapted from William T. Silfvast, *Laser Fundamentals,* Cambridge University Press, Cambridge, 1996, page 438. *Page 187:* UCSB FEL Laboratory, photo courtesy of Jim Allen *Page 191:* William Pelletier Photo Services, Inc. *Page 193:* Douglas L. Peck *Page 194:* Mike Matthews, JILA research team *Page 197:* Lawrence Livermore National Laboratory

Index

LASERS: Harnessing the Atom's Light